The Meaning of Care

The Meaning of Care
The Social Construction of Care for Elderly People

Bernhard Weicht

Leiden University College, The Netherlands

First published 2015 by
PALGRAVE MACMILLAN

Palgrave Macmillan in the UK is an imprint of Macmillan Publishers Limited, registered in England, company number 785998, of Houndmills, Basingstoke, Hampshire RG21 6XS.

Palgrave Macmillan in the US is a division of St Martin's Press LLC, 175 Fifth Avenue, New York, NY 10010.

Palgrave Macmillan is the global academic imprint of the above companies and has companies and representatives throughout the world.

Palgrave® and Macmillan® are registered trademarks in the United States, the United Kingdom, Europe and other countries.

ISBN 978–1–137–27493–9

This book is printed on paper suitable for recycling and made from fully managed and sustained forest sources. Logging, pulping and manufacturing processes are expected to conform to the environmental regulations of the country of origin.

A catalogue record for this book is available from the British Library.

Library of Congress Cataloging-in-Publication Data
Weicht, Bernhard, 1981–
  The meaning of care : the social construction of care for elderly people / Bernhard Weicht.
    pages   cm
  Includes bibliographical references.
  ISBN 978–1–137–27493–9 (hardback)
  1. Older people—Care.   2. Older people—Services for.   I. Title.
  HV1451.W455 2015
  362.6—dc23   2014036799

Contents

Acknowledgements

One of the most important arguments I am trying to make in this book is that the construction of the meaning of care is fundamentally based on and within actual, concrete relationships between people. Similarly, the research and writing of this book would not have been possible without people close to me and is thus heavily influenced by my relations with others.

I want to thank all those who have participated in the various focus group discussions and who have shared their experiences, ideas, opinions and emotions with me and with other participants. Hearing these stories has made me reflect on and rethink my theoretical ideas for this study. I also want to thank those people who helped to arrange and organise the focus groups. The work in this book has furthermore benefited from personal, collegial and institutional support from the University of Nottingham in the United Kingdom and Utrecht University in the Netherlands. Several people have read and commented on different versions of this book, among whom I particularly want to thank Tony Fitzpatrick, Nick Stevenson, Fiona Williams, Saul Becker, Sara de Jong, Christian Karner, Trudie Knijn, Silvia Radicioni, Johanne Søndergaard, Ellen Grootegoed, Marit Hopman and the anonymous referees. I also thank Philippa Grand and her colleagues at Palgrave Macmillan for their support, trust and patience and Gerard Hearne for his quick and efficient editing work. While I take full responsibility for all errors in this book, I am incredibly indebted to the following people who have made useful and challenging comments on the chapters of the final manuscript: Paul Ramskogler, Edith Waltner, Annelieke Driessen, Magali Peyrefitte, Bernhard Forchtner and Joe Greener. Their insightful remarks have greatly increased the quality of the book, and I am particularly glad to be able to call them not only colleagues but also friends.

Writing about care is, like any other practice, impossible without the care of other people. Concrete relationships have given me both the opportunity to write and the ideas, questions and comments from which I could start to develop my own arguments. Chiara Massaroni has accompanied me through the final stages of writing this book. For this and for everything that is to come, I am really grateful. Two people I am particularly indebted to are my parents, Christine and Johann, who made it possible for me to be who I am and to do what I do, and whose behaviour and practices have taught me much of what I know about care.

1
Introduction

Two retired music teachers in their 80s are enjoying their highly cultured life together in their refined apartment in Paris when their world dramatically changes after Anne has a stroke and suffers subsequent progressive dementia. Complying with her demand to promise never to put her into a hospital or care home, her husband George lovingly starts to care for her, protect her and comfort her. Dissatisfied with professional caregivers and disappointed by their daughter and son-in-law's reactions to the situation, the couple gradually turn the apartment into a temporary hospice in which they try carefully to adjust to all the changes and consequences that Anne's illness throws at them. Both George's strength and the couple's interactions become increasingly challenged by Anne's deteriorating health, and eventually George decides to relieve his wife of her suffering by suffocating her with a pillow. This story, beautifully depicted in Michael Haneke's film *Amour* (2012), which, besides other prizes, won an Academy award and a Palme d'Or, identifies several crucial aspects of the meaning of old age, care and dying. The changes in the relationship of George and Anne and the latter's increasing dependence on her husband are powerfully illustrated as essential and inevitable parts of human existence and people's bonds with each other. Watching it in the cinema, I asked myself whether the film could equally be called *Soin* (care) instead of *Amour*. What is the essential difference between love and care, between relating and caring for the other? How much does care depend on loving relationships and do these relationships change during the experience of giving care? The film, which takes place almost entirely in the couple's apartment, also raises questions about home, belonging and feeling safe. Care is ultimately depicted as an inseparable part of the love that two people feel for each other, and at the same time love motivates, guides and restricts George's care practices.

This book engages with these issues, questions and challenges in order to investigate the different aspects, associations and images that constitute the meaning of care for elderly people in society. What role do concrete relationships and concrete places of living play in shaping people's understanding of care? How do people imagine and situate ideal care for themselves, their loved ones and the general public? What are the consequences of the widespread fear of becoming dependent on others, for both carers and care receivers? Do professional carers and/or services bought in the market change the inherent meaning of care? While these are some of the questions that guided the research for this book, they are ultimately also very personal questions. In that sense, this book is also about me. When I started the research and writing process, I gradually realised that the topic of care cannot be studied in a strictly abstract and generalised way. Engaging with issues of relating, belonging and imagining one's ideal life in old age strongly touches one's moral and ethical disposition, ideals and feelings. Does this, however, imply that the meaning of care differs for every individual? In this book, I will try to demonstrate that while care is experienced as something deeply personal, its meaning is constituted by particular societal constructions. Ideologies, ideas and attitudes about care play an important role in defining the situation and people's understanding of giving and receiving care. Additionally, notions of caring change over time (Jamieson, 1998), and Bowlby et al. (2010: 15) rightly state that

> [the] ways in which we experience care reflects our age, gender, ethnicity, health and social status, and will be influenced by our beliefs and values about families and relationships, and hence by where and when we live.

This book mainly focuses on the construction of informal care for elderly people. The term 'informal care' is to some extent problematic, as it might suggest that informal care involves less work than 'formal care'. While recurrently used, there is, ultimately, no accepted definition of the term itself. In my usage, 'informal care' refers to practices of elder care characterised by informal arrangements, personal relationships and intimate bonds. Informal care in that sense is usually unpaid and provided in domestic settings in an unregulated way. Hochschild (2003a: 214) describes care as 'an emotional bond, usually mutual, between the caregiver and the cared-for, a bond in which the caregiver feels responsible for others' well-being and does mental, emotional and physical work in the course of fulfilling that responsibility'. However, these ideal-type characteristics are (often discursive) associations and neither exclusive nor conditional. Even paid arrangements, such as the live-in arrangements of migrant care workers in people's homes, can demonstrate characteristics of informal care (see chapters 2 and 3 and Weicht, 2010). In contemporary Western societies, care is a much debated issue in academia, politics and everyday discourse. Almost everyone will be concerned with care at some point in their life, either as a carer or as someone needing care by others. Yet, care needs are generally seen as an inherently negative aspect of a particular period in someone's life course. The way we think about being old and being in need of care is characterised by anxiety about becoming dependent and having to rely on someone else's commitment. Responsibility for elderly family members or elderly members of the community is a defining feature of how contemporary societies understand processes of ageing, family, social cohesion and mutual duties of dependence and support. Several social analysts claim that we live in a de-traditionalised society (Beck and Beck-Gernsheim, 2001; Giddens, 1998) in which old traditions, structures and authorities make way for new moral questions and answers. This suggests that in all European countries ageing societies and changing family structures require a rethinking of traditional family-based arrangements. Professional care, both provided publically and on the market, has been an important substitute and/or complementary offer to family-based solutions. However, in most settings, informal care arrangements continue to dominate the situation, and in some countries, for example the Netherlands, even a process of de-professionalisation can be noticed and is, not least out of financial considerations, promoted by policy-makers (Da Roit, 2010). Williams (2004), in this context, holds that while the longevity and configuration of relationships have changed, people's sense of commitment has not (for a similar observation, see Fine, 2005). Still, informal arrangements based on family relationships or other close bonds are obviously not the only type of care provision for elderly people. Particularly in the Nordic countries, informal arrangements are accompanied by a diverse and far-reaching net of professional care provision (Pfau-Effinger and Geissler, 2005), while various countries have also experienced the increasing influence of private market arrangements (Meagher and Szebehely, 2013). While professional care arrangements are not the core focus of this book, they will feature in particular in their discursive construction in separation from informal care (see, for example, chapters 2 and 6). Publically provided care or marketised versions of care might challenge traditional meanings of care, and Hochschild (2012) in particular investigates the

Understanding care

Thus, care cannot be reduced to particular practices as reactions to certain needs; rather, care forms a feeling, an identity, a commodity and a way of thinking (Phillips, 2007). Understanding the construction of care helps to understand aspects of people's ideals, motives, attitudes, imaginations, aspirations and desires in life. An impressive insight into carers' motivations and situations in relation to gender is offered by Ungerson's (1987) influential analysis of qualitative interviews with informal carers, in which she describes the process of 'becoming a carer' and the negotiation of this role. She identifies differences in the self-understanding of care between men and women and notes gendered differentiations between the notions of duty and love as the reasons for someone becoming a carer. The self-understandings, motivations, attitudes and ideas that Ungerson describes constitute a discursively constructed moral framework. Hence, an analysis of how we understand and construct care in everyday life, the meaning it has for ourselves, our families, our relationships, identities and our sense and understanding of society and what is right and proper is crucial for an understanding of the possibilities and consequences of any social, political and cultural intervention.

The meaning of care has important gender connotations. Hughes et al. (2005), for example, emphasise care's meaning as both an activity and a culture in order to explain its feminised status and the subordination of carers. A similar argument is presented by Winch (2006: 6–7) who states that carers are 'produced by an interplay of political structures and ethical attitudes and practices', which is based on a carer discourse and a 'morality of caring'. Also, Paoletti (2001, 2002) takes a discursive approach and places care 'as part of the social and moral order' (2002: 815), which is produced and reproduced through ordinary talk. She furthermore argues that the vulnerable situation of carers needs to be explained by the moral context and its gendered nature. Elsewhere, Ungerson (2000) speaks of an ideology of 'natural' traits, practices and identities of women, which 'bear such a close resemblance to the practices based on the experiences of mothering and hence are construed as "natural" aptitudes of women' (2000: 636). Similarly, Guberman et al. (1992) identify 'feelings of closeness and interconnectedness with family, gender-role conditioning, and life situation' as determining a (gendered) caring role. In this context, paid and unpaid care are designed to be based on the specific construction of care 'as a hybrid of love and instrumentality' (Ungerson, 2000: 627).

In order to understand the fundamental meaning of care, I will investigate several aspects of its social construction:

- Firstly, I will explore the moral and ideological underpinnings of the process of caring. What are people's associations with care? What desires, wishes and hopes are related to the social practice of care and its imagination? I will also sketch out the various discourses people refer to in order to organise care for elderly people. What are the possibilities for the contestability of these discourses and their moral constructions with respect to care? Focusing on an enduring paradox, I will investigate how it can be explained that while those who care are continuously valued very highly, care is, at the same time, politically and economically only an issue of marginal concern.
- These questions and issues lead to my second focus of investigation: On the basis of the moral construction of care, what do people define as the 'proper thing to do' (Williams, 2004)? Individuals are confronted with very personal immediate demands in their lives. Care needs do not only require the fulfilment of certain tasks, they also trigger answers in an emotional, intimate way. How are responsibilities, duties and commitments constructed in both family contexts and the broader societal framework? How do carers and non-carers experience and reflect the social discourse on care(rs)?
- Thirdly, picking up Hochschild's (2012) discussion, I want to base care's social and moral construction within current societal developments by asking how care is positioned within a neo-liberal construction of modern society. I want to explore to what extent the concept of care is seen as being contradictory to the economisation of society. I will therefore also challenge popular conceptions of late modernity by authors such as Giddens and Beck. The investigation of the moral conception of care demonstrates the importance of 'traditional' ideals such as family, home and community. Giddens, in his conception of the 'Third Way' (1998: 36) for social democracy, argues for example that the 'new individualism [...] is associated with the retreat of tradition and custom from our lives, a phenomenon involved with the impact of globalization widely conceived rather than just the influence of markets'. He argues that we live in an age of 'moral transition' (Giddens, 1998: 3) in which mutual obligation and individual responsibility become more important features. I will investigate how care is positioned in relation to traditional conceptions of moral living and the developments in and demands of modern society.

For all three interlinked fields of enquiry, it becomes obvious that ethics and morality play an important role in constituting what could be described as the meaning of care. This, however, does not imply the construction of an all-encompassing moral theory; rather, I focus on people's negotiations and ethical existence in everyday life (see Sayer, 2011). People in their daily lives do not rely on grand theories of morality or moral actions but rather apply some form of practical reasoning or judgement, which Sayer (2011) calls 'phronesis'. This practical reasoning is characterised by a concern with concrete objects or particulars and is based on an individual's character, rather than universal rules. The concrete other becomes the focus of people's actions and ethical concerns. Secondly, phronesis includes practical, embodied and tacit or intuitive elements, which focus on ends rather than means. Thirdly, ethical being must be understood as a relationship of interest between individuals and the broader context or, as Sayer (2011: 20) puts it, 'our relation to the world is not merely causal and interpretative, but one of concern'. Care in that sense must be understood as a particular, long-lasting relationship with an interest in the quality of the very relationship and the particular possibilities of flourishing and suffering (Sayer, 2011). Finally, ethical beings act within specific historical, social, political and cultural structures and circumstances, which shape their actions and understandings of their actions. In the following sections of this introduction, I want to sketch out a theoretical framework that tries to capture and elucidate the different elements of this 'everyday morality' that the construction of care is based on. The framework starts from a micro perspective on individual's practices and is eventually based within the larger societal context. Since imagination of the meaning of a moral practice is usually directed at people's behaviour, ideas and motives, the discussion starts with a focus on the ethical individual and their moral motivation. I will then bring the other and people's responsibility in the presence of the other into the equation. As I have pointed out, care is fundamentally based on concrete, particular relationships. I will hence draw on the ethics of care perspective to suggest taking relationships as the starting point for understanding morality. Finally, the moral construction of care and people's sense of the 'proper thing to do' will be based on and within certain societal structures and processes. The social context will thus be understood as a materialistic realm in which discursive and moral practices can take place. With Sayer, my aim can be summarised as being

> to explain social phenomena in a way which acknowledges the importance of social structures and contexts without ignoring their

ethical implications and without denying any role for agency and responsibility.

(Sayer, 2011: 165)

Acting morally

The question of what sort of person one should be in order to act morally is the main concern for Virtue Ethics (Hursthouse, 1999). Virtues, Hursthouse states, are concerned with actions and feelings, which explicitly include emotions as morally significant. Slote (2001: 4) confirms that 'the focus is on the virtuous individual and on those inner traits, dispositions, and motives that qualify her as being virtuous'. It should be noticed that this is based on the idea that doing the right thing does not necessarily mean doing it for the right reason. Agent-based virtues should be characterised as warm (based on compassion and benevolence) and should reflect the person's 'overall morally relevant motivation' (Slote, 2001: 38). An action is then regarded as morally acceptable 'if and only if it comes from good or virtuous motivation involving benevolence or caring (about the well-being of others)' (Slote, 2001: 38).[1] The idea of being virtuous is obviously important for care as it could be argued that caring should be done for the right reasons and being motivated to care for others requires a specific disposition and character traits (see the discussions on informal care in Chapter 6). Darwall (2002) elaborates on the conditions for a person's good or a person's welfare, which cannot be explained rationally but which must be understood as being relative to the particular agent. What a person values is not the same as how much it benefits him or her. What benefits the other would then be rationally desirable for his or her sake.[2] The relationship between care and welfare is one in which somebody who cares desires and pursues this person's welfare. Darwall (2002: 15) furthermore differentiates between empathy and sympathy. While the former is related to respect, which means to take someone's point of view as normative, the latter describes care, interpreted as taking someone's welfare as normative. This distinction, in which the focus is on the treatment of a person according to her welfare rather than her will, results from the conception that, in care, people desire things for a person for that person's sake. In the discourses on care, this idea features as an emphasis of the concrete relationship between the carer and the cared-for and the former's imagined knowledge about the needs and wishes of the latter (see, for example, chapters 2 and 5). Slote (2001) thus argues for an agent-based approach to Virtue Ethics. The other in Slote's conception is seen in a particular way, in that one's 'nearest and

Zygmunt Bauman adopts Levinas' position to some extent, his 'rejection of the typically modern ways of going about its moral problems' (Bauman, 1993: 4), by rejecting normative regulations and universal claims in general. According to Bauman (1993, 1995), everyone is faced with the challenge of responsibility for the other. In that sense, Bauman introduces an interactive element in the sense that 'being for' someone is the basis of individual moral existence. Furthermore, Bauman emphasises that moral phenomena are inherently 'non-rational' contradictory impulses in which people follow 'the habitual and the routine; we behave today the way we behaved yesterday and as people around us go on behaving' (Bauman, 1995: 12). This, in my view, brings together Foucault's conception of the discursive formation (see below) of social actions and Levinas' notion of (unconscious) responsibility.[5] Institutional arrangements, such as the market or state, have, in Bauman's sense, the goal to release individuals from the burden of this personal moral responsibility (1993: 182). Possible tensions arising in the context of institutionalised or professionalised services are explored in chapters 3 and 6. While Virtue Ethics provides an understanding of 'being good' and the importance of a focus on moral character, it remains concentrated on the individual. Bauman's writings help to situate one's moral character within an ethics of relating (acting ethically in this sense means to respond to the demands of responsibility for the other). So now what it means to be an ethical agent could be sketched out, similarly how disposition and virtues form the character of the ethical agent. Bauman's approach adds the potential to see care as an expression of universal responsibilities between individuals and not restricted to particular relationships. This approach, however, misses an important element that constitutes the meaning of care: the centrality of concrete, particular relationships.

Relationships

Morality and ethics play an important role in the context of care in two ways: Firstly, the moral construction of care underlies the practices of care for all those involved in caring relationships. Secondly, care represents a moral ideal in society; it is seen as an idealised form of people relating to each other. The literature has recognised the importance of the moral mindset of those who care, and authors of the ethics of care approach (e.g. Bubeck, 1995; Groenhout, 2004; Held, 1990; Sevenhuijsen, 1998; Tronto, 1993) put care at the centre of the construction of a new morality.

Situations of care needs firstly trigger a focus on the particular relationships within a concrete (family) setting. People with a shared lived experience also share a history of mutual dependencies on each other. The ethics of care approach rejects a construction of human beings as (masculinised) independent actors but takes human relations as its starting point. The theoretical discussion of the ethics of care draws in general on Gilligan's (1982, 1993) work on 'the different voice' in which she identifies two different (gendered) ways of speaking about moral problems, which inevitably include two different modes of describing the relationship between other and self. As girls develop through an attachment to their mother and boys through a separation from her, the former show a somewhat more emphatic individuation than the latter, and, in general, differences in relation and connection to other people occur. Gilligan then notes that separate gendered identities arise and that intimacy and relationships are categories bound more to the female than the masculine identity. As a consequence, girls and women judge themselves in terms of their ability to care (Gilligan, 1982) and they tend to listen to and try to understand the standpoint of voices other than their own. With respect to morality, Gilligan notes:

> Thus it becomes clear why a morality of rights and noninterference may appear frightening to women in its potential justification of indifference and unconcern. At the same time, it becomes clear why, from a male perspective, a morality of responsibility appears inconclusive and diffuse, given its insistent contextual relativism.
>
> (Gilligan, 1982: 22)

Many authors have developed further variations of an ethics of care. Tronto (1993) argues for a questioning of the moral (gendered) dichotomy, which Gilligan identifies but leaves intact, and Held (1990) highlights the historical split between reason and emotions in the history of philosophy and ethics, which, in her opinion, is built on the identification of the human with man, which results not in a universal ethics but in a gendered concept (Held, 1990: 323). Groenhout (2003, 2004), who links the ethics of care approach to a Christian/Jewish tradition of morality (she also mentions Levinas in this context), argues that for an ethics of care a different idea of human nature is needed. Similarly, Kittay (2009) and Noddings (2003) favour rejection of the claim that 'all people are' a certain way and argue more for human relations as a starting point for the development of an ethical position. A care situation – or a potential care situation – taking an ethics of care approach must be understood as a set of relations between people who are interdependent on each other. Groenhout (2003) describes how a focus on interdependence, with recognition of the separateness of everyone, does justice to the fact that humans are social beings. Bubeck (1995) emphasises that this interdependency is not socially caused but humanly necessary, and care in that sense is understood as work that needs to be done because nature makes us dependent on each other (see my description of dependency and independence and their meaning for the understanding of care in Chapter 5). An ethics of care approach also needs to pay attention to feelings and emotions, such as grief, fear, anger, rejection, guilt, shame and aggression (Sevenhuijsen, 1998: 84). This all-encompassing view enables to focus on the moral meaning of concrete relations of dependency. Sevenhuijsen explains that the

> moral repertoire also needs to encompass notions of cooperation, intimacy and trust. Connection, compassion and affectivity should be recognized as important sources of moral reasoning.
>
> (Sevenhuijsen, 1998: 12)

This also means, for further analysis, taking 'epistemological virtues' (Sevenhuijsen, 1998) such as empathy, intuition, compassion, love, relationality and commitment seriously, as ways in which people gain knowledge about the world and themselves. This knowledge derives from those who care and on whom we are dependent, and it is on this awareness that a new notion of the individual citizen and citizenship in general should be based (Sevenhuijsen, 1998). Held (2002b: 20) rightly states that 'care as a value has great potential for recognition as a universal intrinsic value', which does not need any religious or metaphysical presuppositions.

In order to position care in the context of economic, political and social circumstances, questions of inequality, power relations, exploitation and equality need to be taken into consideration. The willingness to care, for example, also requires the ability to care. Financial possibilities, adequate housing and flexibility on the labour market are but some crucial, materialist conditions for the provision of care. Groenhout (2004) states that ethical assumptions often structure political decisions, and the fact that caring is often positively associated with selflessness inevitably has consequences for political and economic processes. Bubeck (2002: 173) argues that the ideal of the 'selfless and self-fulfilled carer [...] is a dangerous fiction that is imposed on women at their own cost'. To uphold care as a moral value and a basis for political achievement of the good society (Tronto, 1993), the concept needs to incorporate the sphere of economics, politics and social status. Lloyd (2004: 248) in this context emphasises that the position of care 'on the periphery of public life' keeps carers outside normal citizenship. Practically, that means that a space for discourse is provided where the carers' expertise and moral understanding are brought in (Sevenhuijsen, 1998), where the notion of the carer, though, is not essentialised and fixed to his or her identity.[6] Similarly, Williams (2001) tries to define an ethics of care as a programme for political intervention and for the construction of a new citizenship, which combines time and space for personal, caring and work practices. But carers do not get political and economic support, which could be associated with their role as people doing important moral and social tasks. On the one hand, caring for someone should be recognised as an inherently moral practice, and it should be valued as such. Depending on and relations with each other are the main culprit of moral practice. On the other hand, carers (and in particular women) should not be exploited and disadvantaged due to this ethics. Bubeck (1995) proposes an interesting way out of the dilemma. The values and virtues of care should not be restricted to a particular person (i.e. a selfless carer) but should become part of citizenship, and care should hence be organised and carried out on a social level (see the discussion on social rights and caring citizenship in Chapter 6 and the epilogue). In an attempt to use this morality for politics, Tronto develops a 'vision for the good society that draws upon feminist sensibilities and upon traditional "women's morality"' (1993: 3). She (Tronto, 1993) states that the way care is currently constructed poses no threat to the moral order and therefore loses its potential for societal change. Also, an emphasis on care always has the danger that care will be romanticised and domesticated (Sevenhuijsen, 1998). For a feminist ethics, 'which aims to make traditional femininity the subject of critical discussion' (Sevenhuijsen, 1998: 61), the link between 'caring values', relationality and gender needs to be dealt with critically. The identification of men with the creation of the human, which takes place in the public realm, and women with the reproduction of the natural and biological in the household (Held, 1990) naturalises the clear separation between the two spheres and so they appear normal and essential to people. An ethics of care should thus not be naturalistic (Held, 2002b)[7]. Larrabee (1993: 4–5) asks critically whether feminists should really refer to an ethics of care based on relatedness and responsiveness to others, and she warns of a focus on 'womanly virtues'.

of injustice and moral disappointment. In order to do justice to everyone in society, people's status needs to be recognised. Normativity and what is seen as right and wrong are not based on a definite sketch of the good society; rather, they are related to the discursive formation of that very society (Honneth, 1995). Honneth argues that self-realisation, a basic and important good, can only be achieved 'when subjects can experience intersubjective recognition not only of their personal autonomy, but of their specific needs and particular capacities as well' (2003: 189). What is seen as right and wrong in society directly impacts on an individual's dignity, as Houston and Dolan point out:

> Yet, in all of this Honneth sees the possibility of value-consensus, of solidarity amongst social groups as to what counts as a laudable characteristic or contribution to the community. Solidarity arises as part of a 'felt concern' for the other's value.
>
> (Houston and Dolan, 2008: 461)

In societal struggles, characteristics and virtues are defined and valued accordingly. The outcome is a creation of a 'moral consensus'[9], which defines the moral order and people's status in society. Fraser (2003a, 2003b) rejects Honneth's claim that subjective experience determines moral positioning in society. In contrast, Fraser's (2003b: 207) approach 'begins [...] with decentred discourses of social criticism', which she calls folk paradigms and which 'mediate moral disagreement and social protest' (2003b: 207), summarised by Fraser as 'moral grammar'. In that sense the moral order in society is not established by a process of intersubjective engagements and struggles between individuals and groups but shaped by the creation of a moral framework. Approaches limited to one perspective, resulting in either a 'vulgar culturalism' or a 'vulgar economism' (Fraser, 2000), need to be avoided. Disagreeing with Honneth's focus on the sphere of cultural recognition, Fraser (2003a) argues for a 'perspectival dualism', which accepts that there is no ontological distinction between the cultural and the economic realm but only a historicised distinction.[10] But how are the 'moral grammar' and moral social order constructed in society? Fraser finds inspiration in Foucault's writings, in that she argues that norms have replaced laws. Individuals in a society are confronted with hierarchies, norms and a social order; they internalise norms and surveil themselves (Fraser, 1989). Power therefore cannot be located in specific sources; rather, social ordering happens largely through individual self-regulation (Fraser, 2008). A good example of this argumentation is Fraser's discussion of gender and needs. Taking a discursive understanding, she argues that 'needs are culturally constructed and discursively interpreted'; this, however, does not mean that 'any need interpretation is as good as any other' (Fraser, 1990: 220). Rather, the interpretation of (in particular women's) needs follows ideological gender-linked dichotomies, such as the dichotomy of home and work (Fraser, 1989). Fraser (1989) rejects a general division between work and family and separation of the public and private spheres. She argues that due to the moral and economic constitution of society these are ideologically perceived as two different spheres. Ideological public–private separations, however, can be found in both the system and the lifeworld.[11] What is needed, therefore, is to

> redress status subordination by deconstructing the symbolic oppositions that underlie currently institutionalized patterns of cultural value. Far from simply raising the self-esteem of the misrecognized, it would destabilize existing status differentiations and change everyone's self-identity.
>
> (Fraser, 2003a: 75)

Deconstruction is therefore a conscious intervention into both the social and the cultural spheres. Fraser (1990) identifies feminism's task as reconciling materialism and culturalism, more generally, and the public and private spheres in particular, for both men and women. Justice and equality are not achieved by simply recognising the disadvantaged and establishing their social identity (as Honneth's work might suggest). Rather, Fraser focuses on the ideology behind categories and their connection to institutions and the capitalist societal order, as can be seen in her arguments in the context of gender inequalities:

> It suggests that an emancipatory transformation of male-dominated, capitalist societies, early and late, requires a transformation of these gendered roles and of the institutions they mediate. As long as the worker and child-rearer roles are constituted as fundamentally incompatible with one another, it will not be possible to universalize either of them to include both genders.
>
> (Fraser, 1989: 128)

Hence, Fraser (2003a, 2003b) explains that the moral framework must be based on a synthesis of recognition and redistribution, with an acknowledgement of political representation (Fraser, 2008). The ethics of care approach has enabled an understanding of relationality and interdependence as the main fabric of society. I have also discussed, however, that values and virtues based on this ethics are strongly gendered and often lead to inequalities and injustices. A challenge to these consequences must acknowledge the way power relations work out in society. When Fraser (1989: 122) argues that 'the modern male-headed nuclear family is a mélange of (normatively secured) consensuality, normativity, and strategicality', it helps to explain how it is possible that care is highly rated and acknowledged but that it coincides with low status and marginal economic and political perception. Because of the internalisation of norms and the 'moral grammar', negative outcomes of the construction of care are not sufficiently challenged. Fraser's focus on 'decentred discourses' (2003b: 207) points to a sphere which ultimately shapes people's ideas about right and wrong. People are inevitably confronted with divergent moral claims and attitudes, but, by drawing on a moral grammar, they know what is socially desirable, expected and the right thing to do. Individuals have to react to different moral demands, which are then judged against a broader everyday morality. When Valverde (2004: 74) argues that people have to 'juggle ethical responsibilities that emanate from diverse sources and elicit heterogeneous responses', it should be added that these responses will be evaluated against the moral grammar present in society. The construction of a common moral framework in society is also indispensable to what Habermas (1989) calls symbolic reproduction. In that sense, societies must reproduce themselves materially and symbolically by establishing group solidarity, socialisation and cultural traditions. Actual ethical practices and general moral norms stand in a dialectical interaction (Van Dijk, 1991) and are historically reproduced. The existence of this continuation enables the persistence of a social order.

In all the different positions outlined above, there is a notion of the construction of values, morals and identities. How relationships are seen and defined, how responsibility for the other is understood or how 'the good character' is constructed constitute both the ethical agent and moral practice. But so far there has been a lack of explanation of the very construction and formation of these categories, ideals and identities. Fraser's account already suggests the realm of discourse as the context in which a moral order is constructed. Recognising the importance of the discursive realm, Fraser (1990) redefines Habermas' (1989) public sphere as a multiplicity of public spheres of discursive relations in which 'public opinion' is created. Fraser also applies her model to care and states that its discursive formation leads to a situation in which all those involved in care are marginalised (Fraser, 2003a). The discursive realm in which the moral order and the moral grammar are created and shaped and in which the 'framing' of in-/justices occurs (Fraser, 2005) is therefore of particular interest to the present analysis.

The discursive construction of meaning

Care for the elderly, which deals with the (care) needs of people in particular settings, (informally or formally) should first be understood as a social practice, that is, an activity exercised in a particular historical, social and moral context. This rather obvious premise leads to the need to define the relation and mutual impact of actual care work, carers and the social and historical situation. Above, I state the significance of everyday morality, a moral grammar and its construction in and through discourse. From my methodological perspective, discourse can be defined as 'an institutionally consolidated concept of speech inasmuch as it determines and consolidates action and thus already exercises power' (Link, cited in Jäger, 2001: 34). Speech in this context and in the understanding of most scholars of Critical Discourse Analysis (CDA) must not be reduced to actual speech acts; rather, it refers to a broad canon of written, spoken and other texts (e.g. pictures, films). Foucault's (1972) concept of discourse, in which knowledge is seen as historically situated and shaped, shows the significance of the sphere of discursive formation and its effect on the individual's everyday practices. Care, like all other social practices, cannot take place in an independent, self-defined and self-determined realm; rather, it is based within a set of discursive practices, that is, various discursive regimes. Like Alexander (2006), I situate the construction of the everyday morality within civil society, a realm that creates moral values and judgements:

> But civil society is not merely an institutional realm. It is also a realm of structured, socially established consciousness, a network of understandings that operates beneath and above explicit institutions and the self-conscious interests of elites.
>
> (Alexander, 1993: 290)

Foucault's (1972) approach to discourse is useful as it enables the rejection of an ahistorical, essentialising theoretical conception (Still, 1994) of social practices. This also excludes a potential naturalisation of individuals (such as carers), practices (caring) or categories (independence). Valverde (2004: 70) in this context argues that '[a]uthenticity [...] is nothing but a culturally specific effect of particular material and

social researchers, and to pursue what the categories of "single mother", "the old", "the disabled", and so on, mean to those who inhabit them' (1996: 68). Characteristics are often ascribed otherwise in policy discourse (Taylor, 1998) and then reproduced by the social researcher. In my empirical research, I have tried not to impose categories onto the participants. I have used the analysis of newspaper articles to identify terms, categories and narratives via which care is written about. Similarly, in focus groups, I have avoided introducing terms and categories myself as far as possible and have tried to draw on people's own terminology and categorisation. Clearly, this implies that the meanings of categories used in public discourse have significance and are related to power relations in the social setting. I have already discussed Fraser's (1989) example of the existence of ideological gender-linked categories such as work/home in this context. Dichotomies are a fundamental and powerful feature of any discursive construction and function as symbolic codes distinguishing civil society's understanding of good and evil or wanted and unwanted (Alexander, 2006). In the following analytical chapters, I will therefore identify the most significant dichotomous structures on which the meaning of care is based.

Dichotomies, such as structure–agency, fact–value, idealism–materialism or subject–object, have a long history in sociology, as both analytical categories and foci of analysis (Jenks, 1998). Durkheim has already emphasised the significance of various dichotomies related to the distinction between the individual and the social, not least in their importance for shaping moral rules (Lukes, 1973). In structuralist theory, as introduced by Ferdinand de Saussure (1972), dichotomies were identified as signs that shape meaning. Importantly, Saussure (1972) argues that in order to make sense in the context of discourse, dichotomous signs depend on each other, for example the notion of 'good' requires a notion of 'evil'. This important and influential insight was taken up by post-structuralist thinkers, most notably Derrida. In his work *Positions* (Derrida, 1981), he argues that while it is true that in dichotomies the meanings of terms depend on each other, these meanings are not abstract or neutral but created within moral and social hierarchies. Dichotomies such as male–female or white–black are shaped within particular power structures through which a hierarchical distinction can be made. Since our discourses are based on the use of the dichotomies, power hierarchies are reproduced.

In the analysis of care, the identification of dichotomies offers many important insights, not least in a feminist critique of gender-linked categorisations. Fine and Glendinning (2005) identify five different streams in academic care discourse, which are often characterised by certain traditional dichotomies. These dichotomies, such as care–work (McKie et al., 2001; Ungerson, 2005), formal–informal and public–private (Fine, 2005) or state–society (Daly, 2002), are extremely important for an understanding of care. However, dichotomies such as public–private do not refer to clearly defined materialist differences but, as Susan Gal suggests, must be understood as a 'communicative phenomenon – a product of semiotic processes' (Gal, 2004: 261), which are bridged and blurred empirically and discursively. In the analysis chapters, I will thus complement the identification of underlying dichotomies with a discussion on the discursive, empirical and theoretical ways of bridging and blurring those very dichotomies. The semiotic approach implies a rejection of the idea that the public and private are certain places, practices or institutions (Gal, 2004). Rather, for Gal, public and private are 'co-constitutive cultural categories' and 'indexical signs that are always relative, dependent for part of their referential meaning on the interactional context' (Gal, 2004: 264–265). I will try to demonstrate that gender-linked, hierarchical dichotomies are and ought to be challenged on all levels; but, at the same time, they remain a crucial component of the construction of meaning. In other words, even if empirically we see instances of the blurring or bridging of dichotomous constructions, these categories should not be ignored as they tell us something significant about the meaning of care and the desires, wishes and fears related to it. Alexander (1993, 2006) in his analysis of civil society emphasises the persistence of a binary code. Dichotomies fundamentally underlie the conceptualisation of different spheres of distinction of groups and communities:

> When they are presented in their simple binary forms, these cultural codes appear merely schematic. In fact, however, they reveal the skeletal structures on which social communities build the familiar stories, the rich narrative forms, that guide their everyday, taken-for-granted political life.
>
> (Alexander, 1993: 294)

Importantly, dichotomies do not only provide the possibility of categorisation and therefore an understanding of moral values in society, they also guide actions and interventions. In her book *The Purchase of Intimacy*, Viviane Zelizer (2005) discusses the ideological, discursive split between intimate relations and economic and financial exchange. The constructed separation of these two spheres, which she describes as 'two hostile worlds', plays an important role in the analysis of care discourses. Importantly, this dichotomy should not be read merely as an analytical method or a theoretical concept, the split between the intimate and the economic underlies our day-to-day understanding of our actions, feelings and beliefs. At the end of her book, Zelizer (2005) states:

> If this book has done its job well, it will help readers recognize what is happening to them in everyday social life. All of us are, after all, constantly negotiating appropriate matches between our intimate relations and crucial economic activities. Choices people make in these regards carry great moral weight and have serious consequences for the viability of their intimate lives.
>
> (Zelizer, 2005: 308)

This book

The theoretical framework presented above shapes the main approach of the research discussed in the following chapters and also generates the very structure of the book itself. In the forthcoming analysis, I attempt to bridge the gap between the theoretical level of broad moral questions and their application in particular situations. What I am sketching out and analysing can be called an everyday morality, which describes how people understand and make sense of their experience of, histories of and emotions about care for the elderly. Each chapter focuses on one aspect that contributes to the construction of the meaning of care. These themes emerged from the empirical research that forms the basis of the discussion (see below). The structure of each chapter reflects the theoretical and discursive conceptualisation of the book: I start the discussion of each topic with a description and analysis of the meaning of each aspect (e.g. relationships) for the construction of care. What are the different aspects and nuances that matter? Why do these particular discursive patterns matter? The discussion of the meaning, associations and imaginations of people will then become more profound by focusing on the different dichotomies on which the discursive construction is founded. As discussed above, dichotomies are essential in moral discourse, but empirically those dichotomies are inevitably blurred and/or bridged. Each chapter will therefore focus on theoretical, practical and discursive attempts to move beyond these dichotomies. In the conclusion of each chapter, I reflect on the essence of the construction of each topic and its contribution to the meaning of care as a whole. I will additionally focus on the progressive possibilities and prospects of each discursive element. If the construction of care is related to a traditional understanding of society, which elements might contribute in what ways to a renegotiation and rethinking of care within a modern, neo-liberal environment?

The empirical study

These discussions will be explored by drawing on empirical studies carried out in different national, cultural and political contexts. Two case studies (carried out in the United Kingdom and Austria) provide both the empirical data supporting the analysis and illustrations for a better understanding of discussions on specific issues. The case studies consist of a discourse analysis carried out between 2006 and 2010 as part of an investigation into the (social) construction of care for the elderly. Both countries are characterised by being part of a European historical development with its moral and philosophical foundation influenced by a Judeo-Christian-Muslim ethical tradition. Furthermore, both countries are operating with a capitalistic, welfare state economy. The United Kingdom and Austria as the exemplifying institutional backgrounds do therefore reflect cases with similar societal structures, which allow the study to treat discursive practices as being based in a similar context. These countries, however, also reflect traditional differences with respect to (welfare) state regimes within the borders sketched out above (see Abrahamson, 1999; Daly and Lewis, 2000 for an analysis with particular emphasis on gender aspects). While the institutional organisation of care is different in the two countries, informal care is extremely important in each one (see also Österle, 2001).

The Austrian care system is strongly based on care provided informally, usually within family settings. In 2011, 442,251 people received cash payments for people with care needs (Statistik Austria, 2014a, 2014b). Eighty per cent of people are cared for at home by close relatives, of whom 80 per cent are women (Österle and Hammer, 2004: 36). Interestingly, men caring for relatives are usually retired, whereas the majority of women caring are under 55 years old. Only between 4 and 5 per cent of people, 65 or older live in institutional settings (retirement homes and nursing homes), and 5 per cent of those who are 65 or older receive some form of formal home help (Österle and Hammer, 2004). This exemplifies the general trend in Austria of 'de-institutionalisation' (Österle and Hammer, 2004). In Austria, long-term care is formally organised by the payment of 'Pflegegeld', a financial benefit based on the hours of care that are considered necessary (see Badelt and Österle, 2001). One explicit goal of the implementation general. I will explore this issue further in Chapter 5, which explicitly discusses the construction of the passive role of care receivers.

The second level of text analysis is referred to as discursive strategies. This can be understood as how specific values, meanings, attitudes and ideas are produced in discursive materials. This, beside other aspects, includes which terms and words are used, for example, whether a specific narrative is being told by using high-value words (words that are generally associated with positive meaning, such as love, compassion, family[12]) or low-value words (such as control, party politics, selfishness, duties, care as a commodity). Which myths, rituals, symbols and pictures are used and created in the text and discourse? These strategies, which can be seen as discursive realisations of attitudes and ideologies, also focus on how actions are talked about. Van Leeuwen (1995), for example, distinguishes between transactive actions (actions through which others are affected) and non-transactive actions. He argues that 'the ability to "transact" requires a certain power, and the greater that power, the greater the range of "goals" that may be affected by an actor's actions' (van Leeuwen, 1995: 90). The distinction between carer and cared-for person is a good example in this context. The way that the actions of a carer are described in discourse is clearly a sign of transactive action, with the ability to affect others, while the cared-for person is constructed as not having the power to engage in transactive actions himself or herself. Additionally, the power of social actors is also emphasised by attributing cognitive rather than affective actions to them (van Leeuwen, 1995). Linguistic relationships form the last level of text and discourse analysis. These can be seen as the consequences of manifestation of the two levels discussed above. Reisigl and Wodak (2001) list several processes that can be identified in discourse, such as exclusion, inclusion, suppression, backgrounding, passivation, categorisation, specification, genericisation, assimilation, collectivisation, aggregation, impersonalisation, abstraction and objectivisation. In other words, this level of analysis tries to identify what the consequences are of the use of certain strategies, terms and narratives.

I link the empirical data generated in the study to various sources and literature covering other countries and contexts. Each chapter can therefore be seen as an empirical investigation covering a particular topic as well as a theoretical, conceptual and analytical step taken to disentangle the different aspects that constitute the meaning of care. Chapter 2 starts with a focus on the construction of care relationships and their significance for the understanding of care. The materials presented describe how relationships in the context of care are strongly defined by values and virtues associated with family care provision. Even though it is obvious that there are also other actors involved in the provision of care, the family still remains the main association. This occurs through an emphasis on values and virtues linked to the family so that family care always becomes the point of comparison. The dichotomies of family carers and professional care workers underlie the construction of an idealised form of care. I argue that the construction of family is a representation of an imagined ideal, which can also be embodied by non-family members. People's homes have a particularly important meaning in this context as the nexus of intimate relationships. In Chapter 3, I will discuss the geographies of care in more detail. I will focus on the utopia of the home and its opposite, the institutional setting. The dichotomy between loving, affectionate caring and professionalised, institutionalised work will be situated in people's understanding of space and places of caring. Why does being in one's own home have such importance in people's care wishes? What is the relationship between the concepts of care and home? In Chapter 4, I discuss the discursive theme of 'community'. Community can be understood as an ideological extension of the family, while the neighbourhood in which a community exists is idealised as an extension of the home. People refer to the ideal of community, and they emphasise the importance of a functioning community for the delivery of ideal care. In this chapter, I will also consider aspects of nostalgia and imagining ideal caring situations. How do people idealise other times and places in order to construct the ideal caring situation? What role does nostalgia play for an understanding of care? Combined with the safe space of the home and the framework of the family, community is constructed as a counterforce to what are perceived as hostile, individualising and pressurising economic, political and social developments.

Having focused on the questions of who, where and how, chapters 5 and 6 will discuss themes that underlie all of the above. Chapter 5 turns to the situation of those being cared for. People express anxieties over dependency and vulnerability when they imagine old age. In particular, I will discuss the construction of a dichotomy of the independent, ideal actor on the one hand and the dependent, vulnerable, elderly care receiver on the other. In this chapter, I also evaluate the consequences of this dichotomy, not only for care but also for social structures in general. Desiring and imagining the ideal of independent living for as long as possible sketches out an ideology that contradicts many values of care. I will highlight the tensions arising from a discourse, which, on the one hand, emphasises close, intimate care, and on the other, idealises independence and independent living. Before concluding the book with some reflections on the meaning of care and its political implications, Chapter 6 takes up the theme of the dichotomies mentioned before and discusses the discursive positioning of care in opposition to markets. In the other chapters, I identify the construction of dichotomies through which care is ideologically and morally positioned in opposition to work, employment, politics, bureaucracy and markets. This is based on a strong aversion to the institutionalisation, marketisation and professionalisation of care. I will combine these themes by presenting what ideal care means and in particular by sketching out the opposite, the creation of a form of care which is undesirable and rejected. I argue that care is not primarily understood as the fulfilment of a set of divided tasks; rather, it is a complex relationship between the person in need of care, the carer and the environment. The carer is referred to not as someone providing certain services, but rather as being the carer. In creating the ideal of the 'pricelessness' of care, the carer is constructed as offering a gift not only to the elderly but also to the society in general.

2
Who Should Care?
The Construction of Caring Relationships

Introduction

Having worked passionately on the topic of care for many years, I often try to imagine myself in the situation of requiring care. What will my own future look like? If I should be fortunate enough to reach old age, will I be able to live on my own until the end? Or, will I have to move into other forms of accommodation with various possibilities for support and care? What will my life look like on an everyday basis? For myself, but also in conversations with others, I ask myself what the ideal situation would be for me? Which picture of living in old age would comfort me not only then but also now? Having thought about these issues repeatedly and having had many discussions and debates with family, friends and colleagues, one strikingly common demand arises: in all the different scenarios imagined, the setting for living and receiving care can change and be adapted to particular circumstances. What is really important for a comforting scenario, however, are the relations with others. Who will be there for me? Who will actually carry out the necessary practical tasks and who will care for me emotionally? How will I relate to my partner, friends, relatives and carers? Who will I relate to on a daily basis? Care itself represents a close and intimate relationship. Particular (imagined) bonds, most importantly family, are at the heart of imagining this very intimate relationship.

The family has always featured very prominently in discussions on care, and close relationships are central to an understanding of what care means for people and society. It is therefore useful to start the analysis of the moral construction of care by deconstructing these relationships: which actors are mentioned and thought about in the context of care? Which subjectivities are presented and how are relationships practices. Thus, what family is normatively imagined to be influences the ideal caring situation. The values emphasised in the discursive construction of family strongly refer to an idea of 'natural' traits, attitudes and opinions. It is not because of abstract principles that one should want to care for one's parents but out of a natural desire to do so. In other words, the ideal of care is care out of love and not the fulfilment of a commitment.

Care as a family issue

Historically, care for elderly people has been provided within and by the (extended) family. Not only did the family unit represent the realm of caring, the provision of care was recognised to be a main duty of (usually female) family members. While the increasing importance of public and private market-organised care arrangements has led to a diversification in the actual provision of care, many associations and idealisations still essentialise family-related values and virtues. In public discourses, the family still holds the main association with care, even though developments and changes are recognised, as in the following extract from a focus group:

> Lisa: For me the best would be to be comfortable within the family bond. However, somehow that's the way things go, it's rather like, that older people are somehow, some kind of shifting off is happening.

In the above-mentioned quote, Lisa expresses two aspects associated with family relations: seeing family as the preferred option, but acknowledging a decline in family commitment. The family is often idealised as the best possible option for the organisation of care, and family relations are concurrently associated with love, dedication and commitment. In the following example from an Austrian newspaper commentary, it is argued that relatives who live up to this idealised conceptualisation should be supported by the state and society:

> One who cares for his relatives at home shows heart with that [...]. And everyone who wants to care within the family should receive help.

> (*Kurier*, 22/11/2006)

Caring relationships in general, and within the family in particular, are crucial to the construction of care and have therefore featured prominently in both sociological and social policy literature for quite some time. I want to point to three exemplary contributions which have influenced much of the thinking about care over the last 25 years. Firstly, in a groundbreaking study of qualitative interviews with family carers, Clare Ungerson (1987) identified a generalised idea of family obligations. Particular normative beliefs about family roles and responsibilities determine the process of negotiation of actual caring roles and care duties. Family bonds in particular are defined by their propensity for care, especially between spouses:

> At an ideological level in our society, marriage is regarded as the supreme caring relationship, rivalled perhaps only by the mother/infant bond.
>
> (Ungerson, 1987: 51)

Ungerson particularly focuses on the process of becoming a carer, which, she argues, is intrinsically linked to taking over a particular role within the family. Especially for women, taking over or adopting a carer role for a family member often means that this 'woman becomes identified as a "carer" for ever and anon' (Ungerson, 1987: 56). Ungerson's description of the gendered assumptions regarding caring roles that underlie family relations is important to understanding the close ideological link between the family and the idealised and imagined relationship of care. In another significant study, Qureshi and Walker (1989) similarly focus on the care relationship but particularly on the experiences of living a care relationship. The authors describe a very close and direct connection between the family and care, and they describe caring relationships between elderly people and their families as 'the bedrock of "community care"' (Qureshi and Walker, 1989: 5–6). Family care, they highlight, bears an ideological function as the simple imagination of family care in someone's life makes people already feel cared for. Qureshi and Walker (1989: 123) further identify an order of preferences of elderly people, listing the people whom they want to care for them, as spouse, daughter, daughter-in-law, son, other relative and non-relative. This clearly gender-linked hierarchy reflects a normative ideal of preferences, which the authors describe as

> a traditional Western normative preference structure. The rules are that close relatives are preferred to more distant ones, any relative is preferred to a non-relative, and female relatives are preferred to male relatives.
>
> (Qureshi and Walker, 1989: 123)

is an important feature of understanding the centrality of the family for the construction of care. Clarke, however, points out that there is also 'evidence that the current generation of elderly people prefer care from independent sources rather than from the family' (1995: 45) and concludes that increasing resources create choices between different options for care provision. It is obviously difficult and hard for relatives to object to the idea of 'natural', affectionate care within the family, and so a decision to choose a nursing home or, in general, institutional, professional care needs to be made against the background of a moral discourse emphasising care within the family and by family members. Importantly, in all of the aforementioned discussions, families are understood as relationally and emotionally laden networks, and not as conglomerations of fixed roles (Fitzpatrick, 2008: 154). In the following focus group extract, 'family commitment' is seen not to be expected anymore, or not as much as it used to be. Interestingly, a clear distinction between responsibility and commitment is formulated, which is partly explained by a change in culture and economic needs and pressures:

> Bea: And so if they're local they can pop in for limited times. And I do think they have a role. I do think that children should be aware of the situation. In my case it was one of my sons who came one time and said: Mum you've had enough. What are you going to do about it? [...]
>
> Fran: But there's less today of an elderly person coming to live with you. That used to happen more.
>
> Bea: Oh yes.
>
> Catherine: Yeah, my granddad came to live with us.
>
> Fran: The family commitment, [...] you cared for your family, you know.
>
> Bea: I don't think that's expected now so much.
>
> Fran: Well, no, it doesn't happen, I know.
>
> Bea: No, it doesn't happen.
>
> Catherine: I think of course it's the culture, isn't it? Two people have to go out to work.

This exchange includes a critical commentary about our modern-day (work) culture and its associated pressures and dynamics. Importantly, the frame of comparison is the natural arrangements, how they *used to be*. As mentioned above, the discursive construction of the notion of family as a loving bond of individuals who are committed to each other often manifests itself as a focus on the 'natural' form of the family (Fineman, 2002), in which family and family care are understood as natural processes, as the short quote below exemplifies:

> Fran: That in a way is almost a natural process, isn't it? Looking after the people of your family.

Ungerson (1987: 129) describes this 'naturalisation' of family involvement in care through which particularly women thought they 'fulfilled their sense of duty to their parents'. Family care in this context is embedded in a specific moral discourse, which refers to an ideology of 'natural' traits, practices and identities of women which 'bear such a close resemblance to the practices based on the experiences of mothering and hence are construed as "natural" aptitudes of women' (Ungerson, 2000: 636). The next focus group extract is an example in which the natural process of family involvement in care is described by the use of references to animals. The process of 'learning' to care and being there for your relatives is described as a natural cycle, which, unfortunately, is disturbed by modern cultural influences and economic pressures.

> Will: We care for our own children [. . .]
> Larry: Look at [. . .] the animals.
> Will: [It's] natural to care for our elderly. But [. . .] in between we get greedy and selfish. [. . .] And other elements of man comes into play and we become, we lose, well, I'm afraid, the present generation of parents, not all by any means, but a lot of children are brought into this world because it's expected. [. . .]. And so this caring element doesn't seem to be quite the same as it used to be, so maybe adults are not learning the skills that they should be, how to care for children and stopping off work, for a few years, I'm not saying whether it's the woman or the man, it could be either [. . .] and working a relationship up, that seems to be getting diluted, and so forth. And that is a worry, that the next generation may not have the care skills, [. . .] they may just not have a clue how to care for their [. . .] parents.

It is important to acknowledge the emphasis on the natural connection between family relations and the care of elderly people in these accounts. The idea of family as a reciprocal relationship is reproduced, but the emphasis is mainly on natural feelings, emotions and love, rather than on considerations of justice or fairness. In a commentary in *The Guardian* on children exploiting their elderly parents, Alexander Chancellor says:

> But it appears that children are the main culprits. How can they be so callous? Their parents are sitting ducks, of course. They tend to trust their children and can't imagine that they would want to do them any harm. [...] It seems incredible that they should allow greed to override their **natural affection** for, and duty of care towards, the men and women who brought them into the world and nurtured them through childhood.
>
> (*The Guardian*, 23/02/2007, emphasis added)

Natural bonds are emphasised and the duty of care is described as being related to natural instincts and feelings. In describing what family means, one needs to be careful to distinguish between descriptive and normative understandings of family. How do people construct a meaning of family which they act upon and how do people think families ought to function? Fitzpatrick (2008) describes two main models for understanding family relations. The 'Indebtedness Model' describes family responsibilities as 'a moral repayment' for care received (seen either in terms of reciprocity or in the sense that emotional bonds established between parents and children create responsibilities and positions of responsible actors in later life). This model has similarities to Finch's (1995) preference for a 'commitment model' in which 'we see responsibilities as commitments which are built up over time between specific individuals' (1995: 54). In opposition to these family-related understandings, the second model Fitzpatrick (2008) discusses is the 'Friendship Model', which focuses more on independent actors. Phillips (2007) additionally points to the family solidarity framework to understand the role families play in care (see also Bengtson and Roberts, 1991; Bengtson et al., 2002). Ungerson (1987: 94) describes family relations as a bond, which is 'based more on willing and highly committed acceptance of an ideology of what family relationships should be like rather than on any particularly strong emotions'. An understanding of care built on trust, commitment, relationships and love can be interpreted as a protective cocoon that facilitates 'ontological security' (Giddens, 1991) and significantly affects the creation of identity. Because people can draw on repertoires of values about care and commitment, worked out through relationships with others (Williams, 2004: 41–42), they gain a secure self-understanding and a conscious (or even proud) position as a carer. The following extract is from a discussion of family responsibilities, a question introduced by me as the facilitator. However, it is important to note that, in this case, the question is about a rather abstract notion of who, generally, bears the main responsibility. It will later be shown that general, abstract principles and ideas about responsibility do not always coincide with decisions, feelings and opinions arising in real life experiences.

> Me: Who do you think has the [...] responsibility to organise minding, care? Is it the family, is it society, is it the person herself?
>
> Ingrid: Yes, principally it's based in the family of course. And [...] that they arrange that with the relative, what she wants, because on that it'd depend, wouldn't it? [...] But, generally of course, the family is the first [...].
>
> Ida: So, especially the children, because at the end of the day the parents have also cared for the children, haven't they? In most cases [laughs] [...]. So therefore one has, I think, indeed a certain responsibility to then also care for the parents.

The main theme raised here is one of filial responsibility, which Finch (1995: 55) describes as 'commitments between a particular child and his or her parent(s) develop[ing] by a complex process based fundamentally on reciprocity'. Finch sees the main aspect of filial piety as making sure that the elderly person stays within the family, an idea that points to the stigma of institutionalisation, which will be discussed in the next chapter. Ivanhoe (2007) describes filial piety as a basic human virtue, which, albeit subject to changes in the cultural, economic and ideological conditions of society, remains based on the special relationship between parents and children:

> While traditional beliefs about filial piety may be out of date, the fact that humans have an enduring, distinctive, and emotionally charged relationship with their parents remains as true today as it was in the past and as true in the West as it is in the East.
>
> (Ivanhoe, 2007: 297)

While abstract notions of responsibility and duty are an important source for the construction of a moral framework, normative ideas about family responsibilities cannot be disentangled easily from particular and concrete relationships. In other words, cultural norms and conventions are both source and consequence of specific situations involving people. A similar argument is made by Finch and Mason (2000: 199) who state that, '[i]n general, people do not seem to "count" the quality of the relationship as a factor which legitimately puts limits upon the obligations of children to their elderly parents'. Limits are rather put in place by other responsibilities held by individuals. There is an important difference, I would argue, between people's general, abstract expectations and their beliefs and personal actions, emotions and opinions. In the following discussion, when Larry raises the question of responsibility for elderly people, Pamela immediately mentions the family as the main unit. Will, however, starts to question this straightforward identification:

> Larry: Who has responsibility? Is it the state, is it the family? Is it community? I don't know.
>
> Pamela: It's the family in England [...]
>
> Will: Again, I sometimes take issue with the family because, I, very briefly, I remember being asked to go down to South-Wales to work, and I said, oh no, I can't, I can't move too far from the East Midlands, because my parents are there and this guy says why? I said because I feel responsible. And he said you didn't ask to be born [...] It was your parents who made the decision, you have no real responsibility. Now, I know society confers responsibility [...] and guilt [...] to look after your parents. And [...] this is quite important because [...] some people are looked at and ostracised because they're not looking after their elderly parents [...]. I would do it in a Christian way but not because they are my parents. I would look after them as I would look after anyone.

This discussion is an example of a very interesting dynamic that can be observed in relation to practical, moral reasoning (Sayer, 2011). Will describes abstract principles of family responsibility and discusses to what extent people have a duty to care for their parents. He also argues that children are expected by society to care for their parents and are denounced if they do not live up to those expectations. The idea that children are not responsible to care for their parents because they were not present as independent actors at the moment of conception is also emphasised by Ivanhoe (2007). Similarly, to his conception of filial piety as the recognition of others as objects of concern, Will himself, on the other hand, says that he would care for his parents, as *I would look after anyone*. Two aspects are particularly important: First, this paradox points to an ideal of care being given out of love, rather than out of duty and principles. Second, people often say they would generally agree with certain principles (e.g. children are not responsible for the care of their parents), though they themselves would not act in that way. By emphasising his own commitment (even though it is doubtful whether, in practice, Will would really care for anyone in the same way he would for his parents), Will positions himself as a good moral being who wants (but is not obliged to) care. This again shows a reoccurrence of the distinction between abstract rules and the emotions of particular relationships. As has been argued, care needs to be understood not as an abstract principle but as a particular relationship between concrete people.

Another frequently mentioned issue in the context of family responsibilities is a feeling of guilt if one does not care or does not care enough for one's relative. In the following extract, Helma talks about the possibilities of arranging a live-in carer for her mother who could take over most of the caring tasks she is performing at the moment, as opposed to having her move into an institutional setting:

Helma: So, I have to say, under certain conditions, I could imagine it with every other person, but not with my own mother. I wouldn't want to do it with my own mother. [...]
Uta: Then you can only put her into a [care] home.
Helma: Yes, I would have to show this strength.
Uta: Would you put her into a care home?
Helma: I would probably [...] until the end of my life have to fight feelings of guilt.

What is striking in this example is the clear focus on the family relationship. While she can, in principle, understand it if relatives do not perform the care work themselves, the situation with her own mother is different. Again a difference is made between general rules about commitment and duty, and personal feelings of obligation (based on love for the particular other). Feelings of guilt are explained as irrational and wrong (as one should not feel a responsibility to care because of general rules), but they occur due to the normative ideals of care out of love and the ideal of loving family relationships. In this context, Bahr and Bahr (2001) even favour care as self-sacrifice and see it as high virtue. They criticise self-sacrifice's negative connotation as self-defeating behaviour and argue that family reality differs from the normative, ideologically based idea of the primacy of individual freedom. Hence, Bahr and Bahr consequently call for a resurgence in 'the sacrifice of self or extensions of self, in the interest of priorities of persons whose needs we see as more pressing than our own' (2001: 1232). They furthermore argue that changes in the conception of care (e.g. making it paid work and/or part of the market) would decrease the element of self-sacrifice in care (2001: 1244) and thus reduce its ethical value. It is important to understand that these ideas can be found in public discourses and people's own experiences. The moral superiority of care out of love (and perhaps even self-sacrifice) constructs an ideal of family care with very demanding connotations.

Labour of love

The conceptualisation of care as a 'labour of love' (first introduced by Hartsock, 1983; Rose, 1983) takes up the ambivalent requirements of and associations with work provided within personal, intimate relationships. Traditionally, families are the context and realm of exchange, reciprocity and affect (Qureshi and Walker, 1989) based on the mutual reinforcement of a combination of personal feelings and normative values (Walker, 1995). Williams (2004) and Twigg (2003) point to the importance of the meaning of people's personal relationships (and their quality) with their children, partners, kin and friends for their very own sense of identity and happiness and the significance of those relations for the quality of care. Picking up on the earlier discussion on indebtedness, love as the basis for caring and reciprocal relationships is, in Ivanhoe's (2007: 304–305) terms, the 'only appropriate response' to the love given by parents, which should result in 'keep[ing] in mind the nature of their love and, in the warmth of this light, to cultivate reciprocal – yet distinctive – feelings for them' (Ivanhoe, 2007: 304–305). These feelings of love as a basis of care relationships can be found in many descriptions of the ideal for society and communal living. Gordon Brown's (then the United Kingdom's Chancellor of the Exchequer) contribution in a newspaper reproduces this strong relationship between care within families, the ideal of love and a model for the whole of society:

> Among the men and women who do so much for Britain are our carers. The six million loved and loving carers of those close to them are the very heart of our compassionate society and an immense force for good.
>
> *(Daily Mail*, 21/02/2007)

Responsibility in the sense of duty is thus more often replaced by a notion of responsibility out of love. Smart (2007) argues that both love and commitment are important for functioning relationships of care.

Separating the two would bear the danger of seeing commitment as good and care out of love as unreliable. She furthermore argues that a

> focus solely on commitment reduces the individual to a one-dimensional being, cognizant only of duty, and it robs the person of precisely the realm of the 'magical' and transformatory which imbues much of daily life with meaning.
>
> (Smart, 2007: 78)

In her interviews, Ungerson (1987) noticed a strong gender difference regarding ideas of responsibility and love as expressed in discussions. Women were much more likely to refer to normative obligations and expectations, while 'the word "duty" was missing from the men carers' vocabulary' (Ungerson, 1987: 92) completely. Men, Ungerson explains, referred rather to love as the reason for their caring. While it has to be noted that these were mainly men caring for their spouse, which inevitably involves certain associations with love, Ungerson's (1987) claim is interesting in that the expression of duty refers to a sense that is 'generalizable from one relationship to another and it is largely unconditional' (1987: 92). Love, on the other hand, is even more bound to a specific relationship between two people and seems to be emphasised much more in the current discursive construction of care relationships. Giddens' notion of a pure relationship (1991: 88), which is characterised by intimate, unconditional love and the absence of economic or other outside interests, might be a useful approximation of the idealised caring relationship, which demands commitment to both the other individual and the social relation itself. Care out of love would be an expression of a relationship between specific others, while a focus on commitment would reproduce abstract, normative rules and rhetoric. Due to the construction of the family as the natural unit and context in which care happens, family care is seen as a very distinct form of care arrangement, specifically in relation to emotional needs, as Marion for example argues:

> Marion: Of course, psychologically, I think it is more ideal if it was the family being there for her [...]. Because [...] the family knows what the elderly person needs. [...] My mother, my grandmother doesn't need to talk. I know what she wants.

This points to a differentiation in forms of care made by Lynch (2007) in which she describes a labour of love as one form which, due to the unavoidable emotional involvement, cannot be substituted by formal arrangements. Because care is associated with love and the idea that 'unconditional love lies at the heart of the family experience' (Kendrick and Robinson, 2002: 294), non-family care arrangements challenge the ideal and the idea of the family itself. If someone resists caring for their own relatives, the family relations between those people are questioned. This also means that professional carers are meant to have a particular attitude, behaviour and identity to be able to provide proper care (Hugman, 2003). The particular construction of care (and its idealisation as a labour of love) plays an important role in the context of gender-linked professions in which it is predominantly women who are employed (Christie, 2006). A focus on vocational attributes and dedication tends to characterise the work performed by professional carers, the majority of whom are women (White, 2002). Muzio and Tomlinson (2012) argue that due to their feminized status, particular professions, such as nursing or care, have failed to achieve full professionalisation and remain gendered semi-professions (see also Henriksson et al., 2006). Similarly, Godden and Helmstadter (2004) emphasise the relationship between concepts such as love and dedication and the professional care sector, pointing out the historically important link between a morally praised profession and the ideology of a 'good woman' (Hallam, 2002). The association of certain professions with vocation has remained a persuasive discursive practice, resulting in a lasting legacy of strong gendered connotations within the caring professions, as Godden and Helmstadter (2004: 174) summarise in the context of nursing:

> The Nightingale construct of nursing was unique in the way it valued gendered, class-based character above professional knowledge, and was to be a lasting legacy. The 'good woman' who fulfilled the woman's mission in ministering, soothing, and comforting, leaving 'strength of understanding' to the male doctors, became identified with the Nightingale nurse. Although Victorian values no longer prevail in society at large, the concept of nurses as comforting and caring women who require little education persist in both the nursing and the wider world today.

Moral ideals and ideas are at the heart of the design of national and cultural care systems, as Kremer (2007), for example, demonstrates. Similarly, Pfau-Effinger (2005), in a cross-national analysis, identified different family values underlying the distribution of care between the public and private sectors and ultimately between men and women. Care, both in its informal form and provided professionally, is, due to its construction, often associated with women (Outshoorn, 2002). Consequently, care's association with the family and values related to the family has particular gendered implications. Fraser, who states that 'affective care is actually women's labour, ideologically mystified and rendered invisible' (2003b: 220), addresses this problem of the marginalisation of care and its reduction to self-sacrifice and moral responsibility. 'As a result', she writes, 'not just women but all low-status groups risk feminization and thus "depreciation"' (Fraser, 2003a: 20).

Dichotomies: The family vs. the professional stranger

As has been argued, the meaning of relationships for the construction of care relies on values and virtues associated with family and other informal bonds. In the discursive construction, dichotomies play an important role in that care as paid labour is positioned in opposition to the very ideal of informal caring relationships. The classical distinction of loving care and professional knowledge-based work reproduces, as de Meis et al. (2007: 328) argue, the separation of the 'house', which is associated with care and the 'street' universe, associated with professional work. As a consequence of the constructed meaning of relationships, this dichotomy manifests itself in two interrelated ways: since family values express some notion of closeness, personal bond or familiarity, family caring relationships are first constructed in opposition to the stranger who is depicted as someone who distorts the realm of family and familiarity (see also the discussion on 'strangers' in people's homes in the next chapter). Secondly, the construction of family care based on love, commitment and responsibility situates caring relationships in opposition to professional care workers, the latter functioning as a representation of rationality and distance. The following discussion exemplifies how relationships with care workers are negotiated in contrast and comparison to family relations. Vera presents a very common view that care workers are, in contrast to family carers, emotionally not involved with the person in need of care. However, Mary disagrees and describes how, for her mother, the situation is clearly different. Interestingly, she uses two main arguments in favour of care workers: the professionalised and therefore more qualified care that is provided and, secondly, the actual inability to distance oneself from the engagement with the people cared for:

> Vera: And with a family member you don't have that, you do have
> emotions and those you take with you, [it's] not possible, you can't

switch that off. [...] And a professional [carer] just doesn't take it with him, must not take it with him, they must not identify with the situation [...].

Mary: No, I don't really think so. Firstly, I hope and I think, that there's a difference of quality, because at the end of the day there's a lot of medical professional knowledge behind it. [...] And secondly, I think, it is not true [...] that you have some distance from it, because I see it with my mother. She's working in care and maybe she should be able to do it, but often it is very difficult to switch off and also to really keep the distance. And she, that's how I experience it with her, she's taking a lot of it into her daily life, that's a big topic in conversations. It comes up, certainly half of the time we're talking is about her work, and about the cases she is dealing with, also in the hospice. So it moves her massively and it also gets to her. So it is not true that you can simply switch off. And I do think that she's doing professional work, and high-quality work.

Vera: Yes of course.

Mary: Insofar I think that it is just a prejudice, and she has indeed learned methods to deal with it, but she's still taking it with her into her life. And I'm not sure how far it gets to her, even more because she's confronted with a number of cases, than for example care in the family, where it is just one case that gets to you. So, I'm not sure, whether it is not a bigger burden.

Vera: Yes, but if that happens to your husband, or to your child, or to your family, with a close family member, you have emotions, I mean memories and [...] a whole life spent with this person. A care worker, even if she is very qualified, they always try to research and investigate the biography of this person, that's clear, but they don't have the experiences with this person at all.

Mary uses two main arguments in favour of professional care workers: firstly, quality of care becomes an issue and professional carers are linked to providing better quality due to education. This education, however, is identified as *medical professional knowledge*, and not necessarily a vocational and/or practical ability to care. At the same time, Mary's account is testimony to the force of the discourse on professionalism, when she states that her mother should perhaps be able to distance herself. Secondly, how the two women talk about professional and family care in this discussion exemplifies the common idea that care is intrinsically based on emotions and feelings of love, closeness and sacrifice. This narrative highlights a number of elements that are based on the underlying binaries in the discursive configurations of care work. Jones and Green (2006) contribute to understanding how this tension is solved by concentrating their discussion on the discursive and rhetorical work that is necessary to allow the combination of both a focus on one's own professional status and an emphasis on the vocational values underlying professional engagement. In order to identify her mother as a (caring) 'carer', 'despite' her professional status, Mary emphasises the fact that she is also emotionally involved in the process of caring. Being touched by it is constructed as a clear sign of 'real, affectionate care', and in order to establish the professional (here her mother) as a carer these characteristics are highlighted.

People often refer to the hard and dedicated work that carers and nurses do. This, however, can never meet the requirements and characteristics associated with care provided out of love and personal relationships. In the discussion, professional carers are often understood as (inevitably) distancing themselves from the people they are in charge of and seen as employees who are simply fulfilling tasks. Failing to respond to individual needs or, more generally, failing to acknowledge individuality, seems to be one of the recurring criticisms of care workers in homes. As the examples below show, however, it is often not the care workers who are accused; they are rather seen as being part of a system that is the problem.

Paul: They can also not deal with everyone there, obviously. [...] I mean, if they once get to know them and if they had the time, then they could...

Ingrid: ...respond to individual needs....

Paul: to individual demands, yes, or rather individual readiness for action of those, who work there. Whereas I don't want to accuse them of something.

Ida: Yes, they do make an effort anyways.

In contrast, Anton and Britta, in a focus group discussion, emphasise the skills and training of professional carers, which gives them an advantage over family members who are emotionally too involved. Here, professionalism is defined as the ability to provide good work because an educated carer can distance herself from the person in need of care.

Anton: Yeah, distance, I think, is a very difficult one. [...] And in the family that's all dependent on one person only. And that's certainly a very big burden for this one person, who also isn't trained for that. Other than in a care home, where there are educated, trained people, who also know what they're doing.

Britta: Who also have the skills!

In both examples, professionalism is contrasted with family-related values and characteristics of care. Whether positively or negatively described, professionalism is a persuasive discursive concept, a set of appealing ideas and ideologies (Evetts, 2003). In the vast literature on professionalism and professionalisation (see Evetts, 2003, 2011), higher educational and training standards for employees as well as bureaucratic procedures for transparency and accountability are mentioned as main features. In the context of caring professions, the concept has gained prominence as a marker of knowledge, skills and distance from the person to be cared for. Discussing the history of nursing, Hallam (2002) shows a development from a quasi-religious to a more secular occupational identity. Similarly, Liaschenko and Peter (2004: 489) argue that, 'historically, nursing has been understood as a calling and a vocation, but modern, secular nursing has essentially understood itself as a profession, and more recently, as a practice'. The desirability of the concept of professionalisation, however, would need to be put into practice within a context in which care is still often defined and framed as a form of 'love labour' (Kendrick and Robinson, 2002) with influential and religiously based roots (Laabs, 2008) and a strong moral demand on its practitioners (Araujo Sartorio and Pavone Zoboli, 2010: 687).

The concept of fractal distinctions (Gal, 2004) provides a fruitful tool to understand the discursive (re-)negotiations of the dichotomy of family and the professional and also within the professional provision of care services. The ideological and moral split between informal, vocational relating with the other and professional and/or commodified work is reproduced in the context of paid care work itself. As a consequence, in the care sector some people are defined as doing professional work, whereas others 'really' care. Claire, when giving an account of her own work as a care assistant, links the ideal care delivered in an institutional setting to the natural aptitudes of people (in this case, not surprisingly, women). She presents foreign nurses as a counterexample of those who are interested in the financial, organisational aspects and who treat employment exclusively as a means to earn money:

Claire: What does a mother need? She doesn't have any education. [...] A bit of common sense and what really needs to go with it is

love. You can really hardly find that anymore in elderly care. [...]
But when I see that we bring over nurses from abroad [...] the first
thought is how many hours can I work, can I work 12 or 14 hours
[...]? How much do I earn? [...] And there I think, where's the
human being in all that? Where's the humanity? What is now in the
foreground? And for me care has indeed [...] drifted apart. We talk
incredibly much, and train incredibly much [...] and document,
yeah, the documentation.

Claire uses a description of very negative stereotypes about her colleagues from abroad to emphasise her own values and virtues in care.
In order to be recognised as 'good' carers, professionals in the care sector thus need to complement the process of professionalisation with
an embodiment and performance of dedication and closeness. Commitment and dedication as values are emphasised, but there is also
a strong notion of *the character* of the caring people, both paid and
unpaid; they are imagined and constructed as a different *kind* of person. So, even those working in professionalised settings can be 'morally
rehabilitated' when their being is constructed to resist the processes
of professionalisation. Personal and moral commitment, involvement
and concern become signifiers of the professional being a carer (see
Husso and Hirvonen, 2012). Practically, the boundaries between dedicated vocational carer and knowledge-based health-care professional are
often translated into sub-positions, such as nurses and assistants (Bach
et al., 2012).

The discursive challenges that arise for professional carers because
of their dichotomous construction similarly apply to those informal
carers who receive financial remuneration for their work. Baldwin (1995)
emphasises that financial remuneration for caring for an elderly person can threaten the carer's standing in the family as it challenges
the norms and values of 'normal' family life and family relationships.
The following extract exemplifies how care can challenge a particular
form of relationship in which a person feels that official recognition as carer (in order to claim benefits for carers) would undermine
family ties:

> Betty: It's the issue that was raised earlier when people are afraid to
> claim benefits. [...] Do I really need it, is it for me to claim the
> benefit, or being called a carer or not while all I do is this, this and
> this, what I normally do, without realising that this is the job of
> a carer? It's the same sort of, it's the cash [...] really, do I claim

benefits, I care for her anyway, or shall I be paid for something I'm
doing all the time, you know?

Nathan: I think too, it puts it in a different relationship. [...]
It changes the relationship. [some agreeing] And, a lot of us would
think it changes the relationship so therefore I wouldn't want to be
classed as a carer.

Clearly, an arrangement in which people are paid for their care work
challenges the notion of a loving family conception. The uneasiness
with the term and the identity of a carer can also be seen in the following example taken from *The Guardian* in which a daughter talks about
her relationship with her mother who is in need of support:

Should I really claim to be her carer? After all, she's not living with me
and I'm not responsible for her every minute of the day [...] Carer
really is too grand a term. What I am engaged in is brinkmanship.

(*The Guardian*, 20/01/2007)

Henderson and Forbat (2002) describe these 'unwanted identities' of
people who do not want to be reduced to an official term. The relationship would be challenged by the inclusion of such an attribution:

The terms care, carer, and caree prevent the construction of assistance
being expressed as a normal component of the relationship. The
terms suggest 'otherness', which places meaning outside of the
interpersonal arena. This highlights a tension between meanings conveyed in policy and those constructed by care participants in their
lives.

(Henderson and Forbat, 2002: 683)

The dichotomous construction of relationships is powerful in that a relationship which is based on family ideas and values would be threatened
if images and narratives of other areas of life were to penetrate it. People
who are asked to combine a role as carer with other activities as a family
member (Twigg, 2003) are confronted with discourses that create caring
either as part of a family identity or as paid work. Zadoroznyi (2009:
280) rightly claims that 'we do not have a "recipe" for a paid "caring
stranger"'. Not only is the carer as a person constructed as someone who
is 'compassionate, emphatic, merciful and selfless' (Winch, 2006: 14),
the doing of care (i.e. care work) is defined as being based on dedication
and, ideally, as has been shown, love.

Beyond the dichotomy

Hence, from the above discussion, we can see that processes of professionalisation in the area of care work have produced a number of new configurations. In practice, an understanding of care as being clearly divided between, on the one hand, informal, family-based care at home and, on the other, commodified, professional care in institutions is misleading and does not reflect the reality of hybrids of love and instrumentality (Ungerson, 2000) and contract and affect (Glucksmann and Lyon, 2006). Imagination and desire linked to the construction of care based on family ideals conflict with a wish for the professional provision of care. Pickard (2010) shows how people involved in care have to juggle two discourses, of an individualised society on the one hand and family ideals on the other. In this section, I want to discuss the consequences of these contrasting discourses, arrangements which constantly blur and negotiate the boundaries of the dichotomy of family and professional care. Taking up the discussion on the gendered construction of this dichotomy, it can be argued that, for women's emancipation, a clear distinction between professional work and informal (family-bound) practices has certainly been important, and Joan Williams (2000) provocatively states that women's problem has been too little commodification, not too much. However, choosing one extreme of the constructed family–professionalisation dichotomy does not often reflect lived-out practices; many of the configurations of care work in professionalised times have a rather more complex form than mere rejection or embracement. A romanticised understanding of family care as a labour of love rather than a professional commitment, and linked to a private morality, carries the potential danger of exploitation. A strictly professional approach to care work, however, ignores the indispensable values associated with caring relationships. Zelizer (2005: 298) thus argues that:

> ... [i]t is not the mingling that should concern us, but how the mingling works. If we get the causal connections wrong, we will obscure the origins of injustice, damage, and danger.

In the context of care, one can see the emergence of hybrid forms and arrangements that try to combine financial transactions with intimate care. It can even be argued that due to economic and social developments a combination of these spheres seems desirable. In fact, markets and state arrangements do play an important role in many societies'

Using the example of the employment of migrant carers by Austrian families, I have demonstrated that the discourse on family care can be extended to other care options as well. It has also been shown that the discourse on migrant carers reproduces the notions of what family is and how care and being there for each other within a family are constructed. In other words, 'family' beyond ties of kin is possible. If the discussion is used to understand what family means for people, it can be argued that family is characterised by notions of love, intimacy and being there for each other, and care within the family is understood as being based on 'affective, quasi-familial, and asymmetrical relations' (Bakan and Stasiuskis, 1997: 10).

The question of what constitutes family care in particular and the ideals of family in general is particularly important for an understanding of other options for care arrangements, besides both professional arrangements and the traditional family setting. Social and economic changes require care and intimacy to be imagined in new, diverse and more flexible (family) settings (see Roseneil and Budgeon, 2004). Heaphy et al. (2004), for example, investigate in this context the non-heterosexual experience of ageing and care. If imagination of the ideal caring relationship is linked to traditional family values and settings, what does the change to traditional settings mean? Since the construction of care is closely linked to the ideal image of the family, '[n]on-heterosexuals have [...] to develop innovative strategies for living outside the normative framework' (Heaphy et al., 2004: 899). If current notions of intimacy and care are not directly linked to a traditional, ideological understanding of family, then how do 'non-normative intimacies' (Roseneil and Budgeon, 2004: 138) define the constitution of family? If family refers to a moral and emotional framework rather than to links between people based on kin and marriage, alternative forms of care are possible. The discussion on fictive kin might provide some more hints for the meaning of idealised caring relationships going beyond the actual nuclear family. Mehta and Thang (2008), in a contribution on Singapore, also challenge the traditional idea of family care, arguing that this image does not reflect the complexities of family care in which also other players, such as domestic workers, exist. Family can thus be understood as an expression of particular values and ideals, which are not necessarily bound to kin ties. Ambivalences can be noticed in people's understanding of family: On the one hand, family values are extended beyond the traditional limits of blood and kin ties'; on the other hand, family represents more than people bearing responsibilities for each other. Rather, relationships tied to care are constituted in a particular way, sketched in relation to the image of family care.

Redefining care – redefining relationships

One part of the caring relationship, those people who provide care and thus attend to dependencies, Kittay (1999) calls 'dependency workers'. One of the main differences between most forms of conventional work and dependency work lies in its relationship to the person in need of care. Whereas, in a Marxist account, one works to sell one's labour, often with no interest in the work as such, in care the interest in the well-being of the other is the main characteristic of the work (Kittay, 1999). The relationship itself is therefore one based on mutual interest and dependence. This clearly excludes the possibility of the dependency worker as the imagined free individual who can be thought of as separate from the person cared for. Eleanor's statement shows the distinction between the 'normal' independent worker and the dependency worker:

> Eleanor: Perhaps one should be looking at how people are going to be encouraged to want to live the sort of life where they are caring for other people, because society now is very much geared to, you are successful if you're making a lot of money and you live in a mansion, your children are going to boarding school and to the continent [...]. In fact most of these places [care homes] run on part-time workers and that of course saves them certain expenses [...]. So probably that's one of the biggest things that needs to be looked at, how to make people attracted to this kind of work and to giving.

The dichotomous construction of loving care and professional work is one of the main challenges for the position of those caring, in both formal and informal contexts. Practices in which the opposition is bridged and in which values of each perspective are combined are rare. One different approach to this challenge is the disentanglement of the different elements that are encompassed by the notion and the ideal of caring relationships. Ungerson (2000) points out that social care, seen 'as a set of tasks', can easily be commodified; the nature of care, however, leads to a situation in which even paid workers are constructed as behaving similarly to informal carers in that they introduce feelings 'in the provision of total care' (2000: 630). I have argued already that the notion of family care should not be literally reduced to family bonds. Rather, family care needs to be seen as an expression of desires, values love and responsibility. The family is thought of as the central realm in which care relationships take place, and it has been shown how values and ideals of care are closely linked to a construction of the family itself. Even though the existence of other actors involved in the provision of care is obvious, the family still remains the main (at least symbolic) association. People seem to be confronted with two opposing cultural discourses: firstly, families are seen as being the ideal care framework; and secondly, care within the family, due to economic and social developments, is not always available and/or desirable anymore. The family as an institution has experienced substantial changes over the last decades, though with no signs of fragmentation (Fitzpatrick, 2008: 143); in the discourse it is still the main association in the context of care. In that sense, all the other options available are judged and evaluated in relation to values such as emotional intimacy, traditionally associated with the family. However, I have also argued that this does not necessarily support a straightforward assumption that family members are seen as being responsible for the provision of care to their elderly relatives. Rather, a complex web of principles, emotions, affections and beliefs, influenced by public, normative ideas, determines individual attitudes and responsibilities. This points to a broader theme, which can be called the relationality of care. The necessary values and ideals, I have argued, can be found beyond the family; in other words, intimacies can be contracted out and employees can potentially be constructed as fictive kin.

The family caring relationship is furthermore constructed in opposition to the professional relation, the latter being based on rationality and contractual arrangements. Importantly, however, this dichotomy is a discursive construction that is blurred and bridged in everyday arrangements of care. Zelizer's (2005) concept of two hostile worlds warns us that the traditional associations with care and the dichotomous associations create problems for the actual practice of care in all the different settings:

> Intimate care sentimentalizes easily, for it calls up all the familiar images of altruism, community, and unstinting, non-commercial commitment. From there it is only a step to a notion of separate spheres of sentiment and rationality, thence to the hostile world's supposition that contact between personal and economic spheres corrupts both of them.
>
> (Zelizer, 2005: 207)

So how do other caring relationships function within this particular historical discursive context? For professional carers, the challenges of the simultaneous associations and demands of a profession and a vocation need to be seen as part of the profession itself (Araujo Sartorio and Pavone Zoboli, 2010: 688). This chapter has also shown that families are the result of moral constructions of values and virtues, and therefore families cannot be seen as 'incarnations of the truth of nature' (Scott, 2004: 231). It can thus be said that the notion of family can be rethought as being defined by certain values, virtues and practices which are strongly related to care. This is not by definition, however, a kin or blood relationship. In practice, care work and the feelings related to it coincide, but in the discourse a separation between the physical work aspects of care, on the one hand, and love for the cared-for, on the other hand, seems to occur. Professionalism and employment do not necessarily mean non-attachment, or an anti-emotional approach, just as 'intimate settings do not stand out from others by the absence of economic activity' (Zelizer, 2005: 291). With reference to Fraser (2003a), Macdonald and Merrill (2002) convincingly argue that (professional) carers need both recognition as altruistic carers and redistribution in the form of better economic remuneration as skilled workers. If professional (economic) exchange and intimate, loving involvement are not understood as contradictions anymore, political interventions can create new arrangements for those doing care work (in a professional or informal capacity). The ideological and discursive distinction between the idea of a morally good person and the economic, professional and instrumental sphere of transactions needs to be questioned. Williams (1989: 75) draws attention to the state's relationship to the family in terms of dependency, domesticity, reproduction and sexuality. Social and public policymaking requires an image of what family constitutes, and Williams (2004: 34) asks whether there is 'a new normative family emerging from law and policy'.

So, if traditional family structures are disappearing, what does that mean for a conceptualisation of care, which is fundamentally based on values related to the very context of families? Are accounts such as Beck's (1998; see also Beck and Beck-Gernsheim, 2001) notion of family as a 'zombie category' and 'Giddens' (1991) idea of a development towards 'pure relationships' what Smart (2007: 20) calls 'a cultural Zeitgeist in which increasing despair about families is on the verge of becoming conventional wisdom'? Along with Smart I would argue that families still have enormous meaning for the arrangement of people's everyday lives and, in particular, for an understanding of care and the construction of care. Clarke (1995) reminds us that families are not stable units and that family means different things to different people, and also lived family practices constitute the idea of what family is (see also Morgan, 1996). Williams (2004) therefore puts effective social practice at the centre of her analysis and focuses on what we *do* rather than on what we *are*. Smart (2007) also adds imagination and memories to a complete picture of what family means. While these critical accounts are certainly true, an element of what family has traditionally meant remains strong and powerful.

This again raises the question of the possibility of other care options, for example, the employment of professional care workers or the commodification of care work. Can real care be performed within a marketised and/or professional setting? Due to the discursive construction, this becomes questionable as Groenhout (2004: 27), for example, warns that '[w]hen a relationship becomes one of rational calculation rather than one of care, the relationship is no longer sustainable'. In Chapter 6 of this book, I will focus on the discursive split of intimacy and the market and its consequences for the construction of care, but at this stage I want to draw attention to Zelizer's argument that traditional proponents of the family model of care use this split against any inclusion of a financial transaction in the realm of care:

> Note that opponents of state-paid family caregiving invoke the now-familiar dual ideas that the intrusion of the marketplace into the sacred space of the family inevitably brings corruption, while introducing sentiment into the workplace reduces efficiency.
>
> (Zelizer, 2005: 171)

Importantly, however, Zelizer argues that the ideological split between care and work has disadvantageous consequences, in particular for women since '[t]urning traditional women's work exclusively into a matter of sentiment dangerously obscures its economic value' (Zelizer, 2005: 89). Similarly, professional care arrangements also need to be understood in their relational focus. Caring for others inevitably establishes a relationship between carer and cared-for, in which the body of the cared-for person functions as both object and subject of care at the same time (Twigg, 1997).

So far, the question *what does care mean?* could be answered by highlighting the focus on the values and virtues of family relations. I have shown that care requires a certain form of relationship, and equally that care produces relationships. Care in that sense is understood as a particular form of relationship in itself. The link between family ideals and care also leads to a situation in which care is constructed in the way family is ideally lived and practised. I have argued that a focus on who owes what to whom, on issues of reciprocity, duty and obligation, misses an essential aspect of the construction of care as 'anti-market' practice. In this form of care, the family is the main realm because the family and one's own home (which I discuss in the next chapter) secure a care relationship protected from the market-logic-based system. Additionally, a focus on individualism and post-familial relations seems to be an overreaction to changing social structures. The values and ideals incorporated and represented by families remain the main feature of what care means to people and what being there for each other means to them. It is not an archaic universality of the family (Burkart, 2002) which makes family such a persistent aspect of discourse; rather, it is the meaning these relationships carry for people. 'Family' in the context of care is not (only) about *who* but about *how* care is thought of. Care, thus far, can be described as the expression of family values, such as intimacy and emotional relation. But care is not restricted to an image of close relationships. In the next chapter, I focus therefore on the specific living arrangements, which build the realm in which care takes place.

3
Where Should Care Be Given and Received?
The Geographies of Care

Introduction

At the moment of writing these lines, I have been living away from my country of birth for almost ten years. Every time I travel back to my home country, I tell people *I'm going home*. After spending a week or two with my family and friends, I again announce that in a couple of days *I'm flying back home*. This story is far from unique. Many people, and not only those with experience of international migration, share these sentiments. When we are asked where we consider home, we often struggle to give a clear answer. *Home is always the place of birth; I can have two/three/four homes; home is where I feel safe and comfortable* are some of the answers with which we try to express the fluidity of the concept of home in our daily lives. However, in all our negotiations with and considerations of the question of which place should be called home, we hardly ever challenge the significance and relevance of the concept of 'home' itself. Even though we might often reject calling one particular house our home, we do not reject the idea of home altogether. Home, in both its macro version of a country or a region and its micro version of a particular house or family, is always constructed as something to be wanted, to be longed for (Duyvendak, 2011). Home has a particularly important meaning in our everyday life. A vast literature on geography has described and evaluated the significance of home and its link to moral and emotional attributes (see Blunt and Dowling, 2006; Haylett, 2003). But for many people home is not exclusively defined by positive experiences. Cold, dirty and crowded places or domestic violence are just two examples in which the home loses its idealised warm and comfortable connotations. Hence, home needs to be understood as an ambiguous and contested concept (Varley, 2008).

In the realm of welfare services, the home fulfils important functions. On the one hand, it represents particular values and ideas and, on the other hand, home is the physical space in which many social policies and interventions, such as care, are carried out. Writers in different fields of geography, sociology and social policy have contributed extensively to the significance of understanding home. While, in human geography, a move towards a cultural analysis of home can be noticed (Conradson, 2003a), welfare geography focuses on the relationship between the moral idealisation of home and the social policies based on this construction (Lawson, 2007; Staeheli and Brown, 2003). In the context of analysing social policies, it is important to note that the discursive construction of home and care in the home finds its continuation in policies and initiatives (Ceci et al., 2012). For long-term care, this means that the construction of home is explicitly referred to in the design of social policies. The relationship between the construction of care and home is thus twofold: First, 'home' should be described as an ideological realm symbolising particular values attached to care through which 'home' is constructed as the quintessential realm of care; as Holstein argues, 'It connotes family, security, comfort, treasured memories, and even "independence"' (cited in Parks, 2002: 11). Secondly, 'home' is a particular physical space that is needed for both living and as a caring space. The home thus fulfils an important function for policymaking around care, both ideologically (as a safe haven and a place in which independent living seems possible) and materialistically (as a place where care is organised and provided and where home care arrangements are enacted). In light of the post-war development of the welfare state and developments, such as ageing societies and the increasing participation of women in the labour market, the landscapes of care have changed (Egdell et al., 2010). However, as care services for the elderly have become more and more institutionalised and professionalised, these tendencies seemed to have stopped, or even reversed. Many countries, such as the Netherlands and the United Kingdom, have furthermore seen a shift away from institutional care back to communities and people's own houses (Egdell et al., 2010).

In public and political discourses, but also in the academic literature, with references to many different national and cultural contexts, it is widely assumed that people generally want to stay at home (e.g. Latimer, 2012). People's expression of a preference for care at home rather than in an institutional setting draws on particular discursive argumentation strategies, which, with Reisigl and Wodak (2001), could be called conclusion rules. The way care is constructed suggests an inevitable, unquestioned and unchallenged rule that *every old person wishes to stay at home* or *old people want to live and be cared for at home*, as exemplified in the following statement in an Austrian newspaper:

Who wants to go into a care home anyways?

(*Kurier*, 23/10/2006)

This construction of institutions as places one desperately wants to avoid is furthermore underpinned by warnings about taking old people out of their familiar environment and *pushing them off* into a home, or even *deporting them* to a care home. The discursive construction of home in relation to care leads to a normative conclusion whereby home care no longer requires any explanation or analysis (Ceci et al., 2012). The unchallenged link between the idea (and ideal) of caring and particular spaces obviously raises questions about the underlying meanings of home for the construction of care in general.

Both on practical and symbolic levels, the daily geographies of care play an important role in people's lives, for both carers and those cared for (Power, 2008; Weicht, 2011). Both physical and psychological restrictions can be increased and eased by the particular arrangements of elderly people's surroundings. Additionally, since care requires direct contact between people (particularly but not exclusively in relation to physical tasks), spatial closeness between those requiring and those providing care plays a significant role (Lawson, 2007; Power, 2008). Geographers therefore focus on the broader locality of the nexus of care (such as rural areas or inner cities; see Parr and Philo, 2003), the communities in which care takes place, the specific institutional context of care (care home, assisted living arrangements, own house), the meanings and characteristics of houses in which care takes place as well as the physical arrangements within the house, including for example access to a toilet and bath (Wiles, 2003).

The home as the nexus of care also has important meaning for gendered connotations of the construction of care. Feminist geographies of home (e.g. McDowell, 1999) have made substantial contributions to our understanding of the meaning of home. Due to its interest in domestic work (including its particular links with capitalism, labour structures and the creation of income) and the structure of families, the analysis of home offers significant insights for the study of care as well. McDowell

(1999) emphasises the centrality of the house as a signifier for both individual status and family relations:

> The house is the site of lived relationships, especially those of kinship and sexuality, and is a key link in the relationship between material culture and sociality: a concrete marker of social position and status.
>
> (McDowell, 1999: 92)

Importantly, both caring as a practice and the private place of the home are associated with women and are constructed as feminised. Building on the discussion in Chapter 2, caring is clearly associated with intimate relationships and the family. This chapter will now focus specifically on the spaces in which these relations can flourish, in particular the home (Parks, 2002). In this chapter, the significance of home as the nexus of intimate relationships (and hence also care) is explored, and its role within the construction of care for elderly people is evaluated. To what extent are the associations related to care, to dedication to each other and to people being there for each other situated within a particular geographical imagination?

First, I explore the meanings of and associations with home and the geographies of care, both theoretically and in relation to the discourses analysed. If the home embodies particular sentiments and associations, how can those be described? In this section, the link between subjective, embodied experience and material conditions as well as social political economies will also be examined (Wiles, 2003). Secondly, starting with the feminist critique of separation of the public and the private, the discursive use of dichotomies in the construction of home will be discussed. In this section, it will be shown that by idealising the home and positioning it as the nexus of care and intimate relations, institutional spaces, that is, care homes, function as the antipode. Starting with Said's (2003) suggestion that an imagination of home also needs an imagination of the other place in order to establish what is not home, discourse is used to demonstrate the creation of a dichotomy of spaces, contrasting the loving, caring space at home and the cold, institutionalised space of the care home. However, the analysis will move beyond the identification of dichotomies by looking at the processes through which the boundaries of public and private, home and care home, and intimate space and public institutional space are blurred and/or bridged. The example of (migrant) care workers living and working in people's private households facilitates challenging fixed discursive boundaries. In these instances, not only people's feelings and associations but also their opinions and arguments show an ambivalent relationship to places in general and home in particular. The final section of this chapter will then raise the question of how far it is possible to separate the values and meanings attached to the discursive construction of home and to use those meanings productively in other settings, beyond the particular place of one's own house. How much is the notion of home intertwined with a particular house (Blunt and Dowling, 2006)? And can this discursive relation be disentangled?

The meaning of home

The construction of the meaning of home certifies it as the ideal place for care. Home plays a crucial role in people's understanding of growing old, of feeling comfortable and safe, and of living an independent life as long as possible. Caroline and Gita's dialogue during a focus group discussion in Austria illustrates the various associations with home and its counterpart, the care home:

> Caroline: I think, that's always the question with a bad conscience. It is, I believe, not only the conscience which says [...] she has raised me, she has always been there for me, she is my mother and now I put her away? And shift her off? [...] they still have somehow a bad conscience then. Plus, there is still the question of would my mother maybe have lived longer if she had stayed at home? In the known environment?
> Gita: Yeah, in a [care] home, my mother wouldn't live any longer.
> Caroline: Because many seal themselves off, they retreat, if they come into a care home.
> Gita: Neither would my aunt. They both wouldn't have lived any longer in a care home.

All these aspects can refer to a particular house, on the one hand, and its broader surroundings, such as the community or a country, on the other (Hanlon et al., 2007). Being at home is for many people the quintessential expression of feeling comfortable, safe and welcome. Milligan (2005) in this context distinguishes how we feel about places and how we feel in places. Both emotional reactions refer to the associations people have with home, a house and a space. Some authors therefore describe home as a site of identity work (Latimer, 2012), where one feels at home and thus also ontologically secure. Easthope (2004) shows that home means places that are inscribed with social, psychological and emotive meanings. The link between home and people's identity is based on both the physical space and an imagined closeness with it, often rooted in people's memories. Bondi (2008) in this context points out that the field of imaginative geographies has specifically drawn on a psychoanalytical tradition in describing the roots of the feeling of home. While the meaning of home can refer to the physical place and the imagined idea of home, the two concepts are intertwined and often related to each other. Even though home is something imagined, conveying particular sentiments and associations, it is also always something concrete and specific, a material entity. Bachelard (1969: 15), in his book *The Poetics of Space*, beautifully describes the relationship between the material house and the ideas associated with home:

> The house we were born in is more than an embodiment of home, it is also an embodiment of dreams. Each one of its nooks and corners was a resting-place for daydreaming.

Describing the interrelation between the imagination of home and the physical place and with the goal of identifying the aspects of home constitutive of the meaning of care, I draw on Marc Augé's (1995) concept of the anthropological place (see also Milligan, 2003, in the context of care homes) for places that convey meaning for the people living in it. Augé emphasises three characteristics of these anthropological places: they entail a certain history, they refer to particular relations to others and they are thus a source of markers of identity. Importantly, Augé points out that anthropological places are constructions, idealisations and imaginations:

> It is only the idea, partially materialized, that the inhabitants have of their relation with the territory, with their families and with others. [...] Nevertheless, it offers and imposes a set of references which may not be quite those of natural harmony or some 'paradise lost', but whose absence, when they disappear, is not easily filled.
>
> (Augé, 1995: 45–46)

However, the possibility of a space being (or becoming) an anthropological space is significant for the meanings attached to it. In a similar way, Duyvendak (2011: 10) asks whether places are particular and therefore 'attachable'. Concrete and particular places can, not least in unstable times, function as refuges and havens to hold on to. In this context, it is interesting that the dynamic instability of neo-liberal reforms in the social sector has led to the increased importance of people's homes, since even in northern European countries, where care for elderly people used to be much more institutionalised and professionalised, the move to increased family responsibility has put people's home back into focus (Ceci et al., 2012).

Similar to the sociological literature's discussion of home as a contested concept (Easthope, 2004), in lay discourses also home conveys various meanings and associations for both individuals and society in general. Duyvendak (2011) describes an interesting paradox in that people usually know what home is and how it feels, and they know *what feeling at home* is; at the same time, however, they cannot describe what *home* really means. Added to that is the fluidity of the meaning of home since the meaning changes according to particular life circumstances, relations and identity struggles. Thien and Hanlon (2009), for example, demonstrate the close associations with an idealised home in rural areas and the changes that the meaning of the countryside (and thus the idea of home in the countryside) has undergone. The concept of home is not fixed, but is in a constant state of re-definition. Authors such as Massey (1995) and Easthope (2004) therefore point to a procedural meaning for places that are constantly in the process of being created and shaped by people. Duyvendak (2011) then distinguishes between different elements entailed in the notion of home, all of which gain significance in different life circumstances. In his account, home firstly represents familiarity with a particular place, reflecting one's history and identity. Secondly, Duyvendak describes the home as a safe haven, which offers protection. Martin-Matthews (2007) in this context refers to home as a relation between territory and boundary and of control and cooperation in which the home provides the power to exclude others and thus functions as a haven and retreat (see also Milligan, 2003). Thirdly, Duyvendak (2011) mentions home's function as a heaven, as a site of positive emotions and associations.

For all these meanings, it must be remembered that the idea of home is referring to both a concrete entity and an abstraction in people's minds (Milligan, 2005). However, the emotional sphere and the materialist world are intrinsically and mutually connected and the meanings of both depend on each other. It thus seems appropriate not to distinguish strictly between home as a physical, material place and an as an idea but, as Blunt and Dowling (2006) do, to try to connect both to the concept of a spatial imaginary. In order to allow this broader understanding of the construction of the home, I want briefly to identify the different aspects that constitute this imaginary. In the following sections, people's abstract associations and feelings will be explored and the home will be situated as a nexus of social relations. First, however, the physical entity itself needs to be discussed.

Materiality

While home relates to many emotive values and ideals, it also gains meaning from the material aspects of the property, the house, and from the possession and use of one's own home. Home thus needs to be understood as a system of boundaries in which one can be in control, in which one can have possessions and where one has the power to exclude others (Martin-Matthews, 2007). Blunt and Dowling (2006) emphasise that physicality/materiality and emotions/feelings always influence each other, that those spheres are not separate but bound to each other. An understanding of the creation of an ideal of home is inherently related to economic possibilities and structures. In the empirical investigation, both focus group and newspaper discourses on care showed that owning a property is significant in several ways: In the discussion about people having to move into an institutional care setting, the notion of having to sell one's property to pay for a care home is often mentioned as an important but negative materialistic consequence. People express, on the surface, repugnance at being forced to sell a property which they have, over many decades, invested their life in. But the aversion goes beyond that. Since the home is closely linked to one's history and identity, in having to sell one's property the psychological aspect of having to give up one's home reinforces and exacerbates the aspect of giving up one's individuality and personal identity. Losing one's home means that one no longer has control over one's surroundings. Milligan (2003; see also Young, 2005a) refers in the following statement on institutional settings to the fact that losing one's private home also means losing the power related to home:

> Institutional spaces are not seen as belonging to residents but to staff, with significant areas of the home 'off-limits' to residents and their families who, in turn, have limited ability to establish spatial exclusion. The application of the term 'home' is thus something of a misnomer and the power balance exhibited within the private space of the domestic home is reversed.
>
> (Milligan, 2003: 466)

In other words, the possibility to exclude others and protect one's *own* sphere is inherent to the idea of owning a place. The significance of owning property can also be observed in the concept of 'housing careers'. The economic significance of housing relates to a feeling of (economic) safety and security, which strongly reminds us of the associations with home (Blunt and Dowling, 2006). And the socio-economic context is bound to and influenced by the economic circumstances underlying care. Phillips (2007: 117) shows the impact of class or educational level on the access to home care. She demonstrates that the higher the education, the greater is the geographical distance between parents and children, which inevitably influences the possibilities to provide and/or arrange care at home. In these situations, the idea of home as a luxury, in the sense that only those with sufficient funds can afford the positive associations with 'home', is at least partly challenged. Pink (2004) rightly emphasises that besides material objects home is also linked to a specific embodied experience of a certain place. The materiality of home includes not only the physicality of the house and the objects and arrangements within it but also other sensory experiences associated with a particular place, such as touch, smell, sound and vision, which all express an embodied understanding of home, knowledge and practices of home (Pink, 2004). Later in this chapter, I will show how these sensual relations to the physical space play out in the construction of binaries between home and the institutional context.

Associations and feeling

Emotions and feelings are often linked to particular spaces and thus engrave those places with meaning and attachment (Milligan, 2005). The physical space of one's own house is ascribed with meaning through images, dreams and imagination that arise through the lived experience of a place (Bachelard, 1969). Home as an idea but also as a physical entity represents 'warm' feelings and positive associations with family, love and comfort. Parks (2002: 11) thus points out that 'home care symbolises all the positive associations we have with hearth and home', and the home offers a space in which a ' "kinder, gentler" form of health care' can be delivered and in which people 'can remain in the bosom of their family'. Images of being cared *for* and being cared *about* reflect similar sentiments and feelings, and as such home becomes a central feature of positive imaginations of care. The following example from a focus group discussion held in Austria shows the interrelation between the physical space of the home and feelings of safety and care associated with it:

> I: I'd like to briefly talk about care at home, [...], that's what all of you have said actually, that it is better for those who are cared for

[...] or it's easier, or more comfortable, if that happens at home [...].

Barbara: Yes, the familiar environment. [...]

I: Why is it nicer, better at home?

Adam: Because the cared-for person has probably lived there for 20, 30 or 50 years, in this environment, knows the people [...] and if he's now going to move somewhere else he doesn't know anyone, he doesn't know how mean the people are. He doesn't know the house customs, if he's getting away from home in the first place, he has no support at all, he's on his own, and he's dependent on them, on the whole system, dependent on the carers.

Barbara: Then, additionally, the old people very often start being demented. They then don't have a proper sense of orientation; [...] they don't know anymore where their things are. At home, they know exactly, this I have here, that I have there, and this they forget again in a new environment, if they have dementia for example.

As will be discussed in more detail below, a dichotomy is created in which the home is associated with comfort, safety and feeling at home, whereas the institutional space is thought of as potentially hostile, mean and lonely. When Adam argues that, in a care home, people are on their own and have to manage alone without any help, the experience of an individual within the space of a care home is constructed as secluded, lost and isolated. One's own home's familiarity, on the other hand, gives people peace of mind (Duyvendak, 2011) and a feeling of security. The feature of home as providing the story of one's life is a very prominent association in the discourse. Young (2005a, 2005b) describes the home as an extension of bodily habits, particularly for older people who are marginalised in society and whose expression of their own identity is somewhat limited to their own home. Similarly, Milligan (2003) sees the home as an embodiment of identity and self-expression, as an anthropological space, which therefore provides security and safety. Only in one's own home can one be oneself (Conradson, 2003b) and retain one's identity:

> As the embodiment of identity, the anthropological place of the home further places limits on the extent to which an individual can be objectified and depersonalized – stripped of their history and identity – to become anonymized within a collective (institutional) regime.
>
> (Milligan, 2003: 462)

The home functions as a display of the personal life and personal identity (Williams, 2002). This feature of an anthropological space, in which people are, by definition (of home), independent subjects within a familiar environment, feeds into the association between family and home. Furthermore, these imaginings and discursive constructions feed into an understanding as if the physicality of the home itself was already healing or caring (Ceci et al., 2012). Because of these associations, 'home care' is constructed as the logical, human and humanitarian option, rather than simply being a cheaper solution for care provision (Russell, 2007).

Social relations

Thirdly, the home gains meaning through the social relations on which it is built and which are performed within it. As has been shown in Chapter 2, close relationships are constitutive of an understanding of good care and the home is a main site for family practices (Yantzi and Rosenberg, 2008). The home plays an important role in constituting social relations, and its meaning is, at the same time, strongly shaped by those relations (Blunt and Dowling, 2006; Staeheli and Brown, 2003). Massey (1995) shows how different locations are closely linked to particular sets of social relations. Care is intrinsically a social activity, and the kind of care relationship therefore shapes the sites in which it is provided and received (Conradson, 2003b; Milligan, 2005; Parr and Philo, 2003). However, the spatial manifestation of social relations is affected by the same ambivalence as relationships. On the one hand, the home embodies the ideals of family care and thus of being embedded within specific social relations; on the other hand, the home is seen as a site of preserving one's individual identity and independence. Relationships acted out within the space of home are seen as underlying one's identity. Understanding home as a relational space is also linked to an association with home as a therapeutic space (Williams, 2002). Williams (2002), exploring the relationship between place and health, argues that the meaning of home is based on it being the realm of relationships with concrete others. The home, with all its aforementioned attributes and associated feelings, well summarised by Young's (2005b) idea of dwelling, is the physical manifestation of what is commonly associated with family life and being comforted by a familial environment. Mallett (2004: 63) argues that home 'locates lived time and space, particularly intimate familial time and space'. The importance attached to one's own home has consequences for family members and often results in feelings of responsibility and guilt. People often express that securing care for their elderly relative in their own home (or in a relative's home) would for them mean fulfilling their duty as a relative (see the discourses on the employment of migrant carers below). This relational meaning of space leads to home being seen as healing and caring in itself. In public discourses, the home is often used as a manifestation of care and close relations, as can be shown in Ingrid's statement in the following quote from a focus group:

> I: And what was the motive, to do it like that [care at home], and not [...] a care home?
>
> Ingrid: Because when my mother was still doing well, I promised her to never put her into a home. [...] And this promise I've kept.

In that sense, care is not something that is added to the experience of home; rather, the home itself is constructed as an expression of care and relations (Latimer, 2012). The idea of the place and space in which care occurs is of utmost importance in the context of discussion of possible solutions and alternatives to family care (see also Mallett, 2004). However, since relationships are a crucial aspect forming the meaning of home, the ideology of home also needs to be understood within the expression of power relations (Brown, 2003; Milligan, 2005). Being in charge of one's own life and having the power to make decisions about one's own life is one of the main goals of living independently and living at home (Ryburn et al., 2009). Kontos argues that one particular feature of home is that it, 'unlike many other accommodation options available to frail older people, does not compromise their independence' (1998: 168). Home can therefore be seen, by definition, as enabling and guaranteeing independence for those living in it. The home, in contrast to an institutionalised, other-controlled existence, entails the possibility of independence and independent living. Maria, in an interview on her understanding of care, describes what independent living means for her:

> For me it means living in one's own home surrounded by one's own familiar belongings and lifestyle. It means privacy. It means still being able to get out to theatre/cinema/social gatherings if desired.

Being able to stay in one's own home is thus associated with a particular level of independence and, simultaneously, the possibility of (freely chosen) social relations. Even in the context of palliative care, home can fulfil this idealised function of combining the aims of independence and family embeddedness to some extent (Wong and Ussher,

2009). Even though people are living in their own home, being cared for by family members and visited by carers, or even living with a live-in carer (such as in the case of the migrant carers in Austria), they are not necessarily 'more independent' than people governed by institutional arrangements. The connotation of 'home', however, secures this experience of independence for people. Kontos similarly states that home is associated with independence,

> by defining a space that is controlled by and is uniquely the domain of the individual. Home is a space in which to pursue personal interests and also, as it is resonant with experiences and expectations, it is a vital facet of self-identity.
>
> (Kontos, 1998: 189)

Dichotomies: Home vs. not home

Also, in the context of geographies of care, the construction of meaning is established by differentiation, that is, the meaning of home is (also) defined by the construction of what is *not* home. A clearly demarcated dichotomy is created between the home and the institution, in which the care home functions as the quintessential antipode to loving care and feeling at home. The institution, described as a cold, lonely and negative place, symbolises the antipode to caring communities (Duyvendak, 2011). In this section, I will focus on the implications of this dichotomy in all its variations. To what extent is the home constructed as a 'haven' in this public sphere? And what does that imply for our understanding of care?

The dichotomous construction of home and the institution has important gendered assumptions. Feminist perspectives have established the significance of differentiation of the private and public spheres in which the private home is associated with the feminine (Held, 1990; McDowell, 1999; Pink, 2004) and work within the home as women's tasks (Oakley, 1974). Similarly, care for elderly people is discursively based within the gendered construction of the home. This projected association between home and the idea of care as a family issue particularly defines women's role differently to men's. Family relations are still seen as the most protected form of privacy, far from the public world of markets, bureaucracy, politics and paid employment. The emphasis on the importance of home thus has a particular gendered connotation of family relations and values:

Caring is tied not only to women but to the private sphere where intimate relationships flourish. This is primarily the sphere of the home and family. Since women have been linked historically to the private sphere of the home, the task of caring again comes full circle to an association with women. And women internalize this association with caring such that feelings of guilt arise if they are charged with not caring enough or, worse, not caring at all.

(Parks, 2002: 21)

By reconstructing the realm of home as a sphere of comfort, security and familiarity, family members and, especially, women are put in a vulnerable situation. Bearing the responsibility for providing a 'home' for their relatives becomes a task that is not limited to practical arrangements but more and more based on a symbolisation of home. The own wish and own position within family relations are then partly defined by the availability of home for frail elderly people. Exemplifying the gendered associations with the arrangement of home, in the following extract Vanessa links the importance of securing a home for her mother explicitly to her identity as a woman:

Vanessa: And again, it was the men who were rather in favour of a [care] home. [...] And we've seen, however, that it goes relatively quickly and therefore I said no way, there's no question about it. We give her the time at home. [...] And she also wished that very much, to be at home.

The construction of the other: The care home

Taking up Said's idea (2003) that an imagination of home also needs an imagination of other places in order to establish what is *not* home, institutions such as care homes play a crucial role in defining the meaning of a home. Institutional care places are constructed as the antipode to loving, affectionate care that features prominently in people's stories and narratives. What real care is must be understood by grasping how care homes are constructed. Similarly to the discussion of home, care homes feature very prominently in people's ideas, stories and narratives about care and old age. However, rather than seeing them exclusively in the context of particular, personal experiences, care homes need to be understood as a concept, symbolising an unloving space, a space that is not home. In discourses, people frequently tell stories and anecdotes about the life and situations in care homes and demonstrate knowledge of the legal and political circumstances. These anecdotes and narratives can usually be characterised by negative connotations, emotions and opinions. The concept of a care home stands for an institutionalised, professionalised and de-personalised form of living, and hence is constructed as a counterexample to dignified living and loving care. In an era of de-institutionalisation (Duyvendak, 2011) in which care institutions are seen critically and more and more services are moved (back) into the community, the care home still has important stigmatic connotations, which are used to demarcate the meaning of home and care itself. In a focus group, Caroline and Brenda discuss the experiences of people at home when they describe the shortcomings of institutional care arrangements. In the home, they argue, care can really be experienced, even by demented, and possibly even paralysed, people. In other words, people can still experience and sense home (see the discussion on the sensory experience of home above). In the care home, however, this experience is taken away and cannot possibly be provided:

> Caroline: The environment, she is at home in a house, really normal. Even though she can't, in a sense, express it, realise it, but the feeling is there. The family, the familial situation there [...], you do feel that, she does indeed sense it. Even though she can't [...] express it anymore.
> Brenda: Yes, you're feeling that through all [the] senses.
> Caroline: Exactly! And in a care home this is, of course, gone.
> Brenda: Whether she's smelling it, or seeing, you do sense that indeed. [...]
> Caroline: Exactly, exactly! In the care home this is of course gone, it's a cold environment.

The care home is sketched out as a non-place, which, in Marc Augé's (1995: 90) words, 'does not contain any organic society'. Furthermore, the discursive construction of the care home already points to an ideological aversion against professionalisation and institutionalisation, an aspect that will be discussed more prominently in Chapter 6. The broader dichotomy is reproduced in the construction of home and care home as two opposing symbols. The archetypical opposite of independent, self-determined living is the institutional arrangement. The care home symbolises everything that challenges a good and fulfilled life, and by using this symbol people can express their fears, worries and negative feelings about old age, being frail and needing care. People in focus groups often argue that in *care homes dementia is rising,* that *people are closing off* and that they are forced into an *unknown environment,* an *alien environment* with *alien people* where there is *no individuality* and *no dignity.* Whereas one's own home secures individualisation, individual meaning and personal identity (Young, 2005a), the care home is a 'placeless space' (Twigg, 2000b), a site that lacks anthropological meaning for those within it. Milligan (2003) uses Marc Augé's (1995) concept of non-spaces to describe the institution as a site in which personal histories, narratives, feelings and identities are absent. The repeatedly used expression that the *care home is definitely more impersonal* refers to this lack of subjectivity and individuality. The possibility of dignified living in a care home is then often rejected and denied. Ingrid, in one of the focus group discussions, describes the situation when a family member moves into a care home as a *very brutal solution* and that people do not want to *shift off* their relatives:

> Ingrid: That's the bad thing, I think. In the home they not only take their [. . .] individuality, they also take their dignity.

This absence of individuality is one of the key markers of the construction of the care home. This conception reminds us of Augé's (1995: 83) description of non-places where everyone has to live according to 'the same code as others, receives the same messages, responds to the same entreaties. The space of non-place creates neither a singular identity nor relations; only solitude and similitude'. The care home signifies a loss of individual identity (Varley, 2008), a loss of close relations, and a loss of memory and attachment – the care home signifies a loss of home. In the discourses, the care home means the absence of care as an expression of significant relationships. The following newspaper extract from an article entitled *'would never put him into a care home'* about a woman caring for her husband at home, illustrates that. The relationship is defined by their mutual aversion to care and living in an institutional space, the care home.

> She employed a nurse for three nights. For 211 Euros. 'But that drove me mad', her husband says with a soft voice. 'You are shepherded that much and still you feel constricted if someone comes after you even to the loo.' (. . .) As long as it is medically justifiable her husband should stay in the environment he is used to. 'That's the only thing I can give him in this situation', she says. 'I would never put him in a

home. With this illness it is so important to have somebody around. Those who are alone don't have a chance to get better.'

(*Kurier*, 13/08/2006)

The loss of individual identity and the absence of significant relationships are often discursively linked to maltreatment within institutional settings. While neglect and abuse of elderly people occur in both private homes and institutions, these topics are regular themes appearing in newspaper articles mainly in relation to care homes. By making it the primary narrative, a particular link is established between care homes and the occurrence of these practices. The following examples chosen from several UK newspaper headlines reflect the relevance of these themes:

Half a million old folk are mistreated says charity

(*Daily Mail*, 06/02/2007)

Elderly 'need new rights to protect against abuse'

(*The Times*, 15/06/2007)

Catalogue of abuse in NHS care homes

(*The Guardian*, 17/01/2007)

Damning report highlights 'harsh reality' of care for elderly

(*Daily Mail*, 10/01/2007)

Some newspapers (e.g. *The Sun*) focus particularly on these issues and care is mentioned mainly in this context. And mistreatment and awful living conditions are continually recurring themes in the discussions. The first associations with care homes are often stories, experiences and emotions related to maltreatment. Many people do have their own experiences of or know stories about problematic practices in care homes, and a general anger with care homes can be noticed. Using a frequently recurring image in a focus group, Brenda for example speaks about the experience of carers in homes not taking the time for patients, explaining that they *put the food there and cleaned it up again. Useless, because she couldn't eat anything anyways.*

Brenda calls this an *eat or die* practice, and this kind of problem is mentioned in many different contexts. Similarly, Peter tells a story about a care home that a friend of his went into. In his account, some classical aspects of neglect or maltreatment appear, such as the massive rooms and the television constantly being switched on:

Peter: [...] the difference between care homes [...] the one that Lucy was in [...] it was horrendous [...] and Lillie quite openly said if she'd have to stay in there she would have committed suicide. Even though she didn't believe in suicide but this was a sort of place where there was a massive room, [...] sort of set off into smaller places by bookshelves but [there] would be a television in each section and they'd be on different programmes, all very loud.

Another symbol for the unattractiveness and undesirability of care homes is stories and images of smelly, cold and dirty places. John's statement can be taken as an example of the associations that many people share about care homes:

John: And then slowly, you start getting a picture. I suppose if I go into a care home and if I – if I smell, unfortunately can't but say urine, that immediately puts me off [...] if I go into that place.

Very often, a strong focus on the costs of care homes and the difficulties in meeting these costs leads to a separation felt on economic grounds. Living in a true home (in the symbolic, ideological sense) is hence seen as a luxury, and only those who can afford it can get a real 'home' (Young, 2005a). A reoccurring image related to care homes, which could be said to symbolise the 'coldness' of care homes, is mentioned by John in his account of the care home his mother was in. Care homes are described as marketised, profit-seeking institutions – characteristics that do not relate smoothly to the demands people have of care arrangements.

John: But I do find that an ideal care home should be a place of security. An environment, clean and properly staffed. Also, there are the social implications of it. Particularly regarding my Mum because the first thing the social worker said to Mum, who was greatly ill, [...] 'you've got to sell your house. Have you got your own property? You've got to sell it.' So in one sense there's not just, there is the financial aspect of it.

The issue of the care home being an institutionalised and therefore also a marketised space is furthermore stressed in a discussion in which the cold, bureaucratic working styles of care homes are emphasised. People working in care homes and the institutions themselves cannot act like they would at 'home' because they are part of the economic sphere of life. The bureaucracy, administration and working arrangements are also examples and illustration of the fact that a care home is not home in the ideological sense. People employed as carers in institutional settings face a situation in which they are working in paradoxical surroundings; for, on the one hand, their work space is constructed as the antipode to home and, on the other, the ideal of their work is seen as providing 'homely' care. The commonly held understanding of care can, almost by definition, not be offered in a care home. Also, social services, care services or other forms of mobile intervention can only meet demands insufficiently. Because care in one's own home is constructed the way it is, and because of its definition as being more than completing particular tasks, formal arrangements must be experienced as disappointing. Zechner and Sointu (2008) describe how formal services in people's own homes are used as 'medicine against loneliness', but it also seems inevitable that these are not equipped or prepared to fulfil this anticipated role. Similarly, in care homes, carers cannot offer what is essential and necessary to fulfil a particular ideal of care. Lack of staff and restricted availability of funding are reasons commonly mentioned as to why carers in homes can only do physically necessary tasks. They cannot, however, 'really care'. Being at home implicitly means a blurring of the boundaries between medical or nursing tasks and personal attention. To get some notion of 'real' care in nursing homes, care workers would need to give additional concern and work with the ideal of selflessness. In the following extract, the discussion revolves again around this problem, the impossibility of the right attention in care homes and potential ways to bridge the separation between home and care home.

Adam: And there's also something else with the care home. I was there, and I've also been told, in [town] there's an old people's home next to the hospital [...] A former hospital [...]. Yes, and they're sitting there, if there are no additional, volunteering supporters, they sit in their rooms and whine, [...] don't mind whether or not they have eaten anything the whole day.
Walter: Yes.
Adam: That's also not care, is it? Clearly, the one person, or the two women, or men, who are doing the care there, they also can't care for 41 or 50 old people.
Walter: That's right.
Adam: That's the [...] problem. And that's again, as you're saying, that's what's missing.

Walter: It's a question of money [...]. If nobody cares then you'll have the poor old lady [...] sitting with her spinach and can't move the spoon.

Adam: That's it.

Walter: And what's missing, you always say, [...] idealism, of course, that's just an employee there. You mustn't forget that.

The home as the realm of domesticity, idealism, selflessness, love and intimacy, in other words the anthropological space 'that creates the organically social' (Augé, 1995: 76), is opposed by the sphere of professionalisation, marketisation and bureaucratisation, the non-place which creates 'solitary contractuality' (ibid.). With Twigg (1997: 228), therefore, home can be defined as 'a secure haven against the hostile world of work'. Bowlby et al. (2010) use the concept of ontological security (see also Giddens, 1991; Shilling, 1997) in this context to describe care's potential to give security and stability to life's course. The construction of home as a family refuge, embedded with safety, comfort and individuality, provides the means for an imagined steadiness of one's own identity, outlook and purpose in life.

Beyond the dichotomy

I have argued so far that the dichotomies of place and non-place, home and not home, own house and institution are constitutive of the construction of the meaning of home, and these subsequently influence and describe the meaning of care. I have already pointed out that the construction of home as a traditional private space entails an implicit reference to traditional gender constructions. Young (2005b), for example, focuses in her discussion on home on preservation, 'a typically feminine activity'. She argues that preservation, in contrast to the sphere of markets, politics and industry, is traditionally women's work and as such devalued and unrecognised. The dualistic construction, prominent in Western thought, shapes the domestic sphere in light of the other, the feminine, the opposite to the public sphere (Varley, 2008). In the public discourse, the link between women and the realm of home becomes obvious. Hardly any person specifically mentions women as those who should do domestic work and stay in the private sphere; the construction of home care, however, shows clear gender connotations. In the following quote, Claire refers to the care that has been lost in modern society. In both child care and care for elderly people, it is women's involvement in the labour market that prevents real care in one's own home.

> Claire: The women go with the first bus to [the city] to the [super-market], at 5 in the morning, come home with the last bus, the old person has to be shifted off to the day care centre, or into a care home, the children have to be shifted off to nursery, and then?

In Chapter 4, I will discuss the nostalgic connotations of these ideas further, but I first want to point out how, through the ideological construction of care at home, women in particular are vulnerable or marginalised. The dichotomies of public and private, home and not home are discursive, ideological constructions, which are constantly questioned, challenged, bridged and renegotiated. The dichotomy of home and not home does not apply to everyone in a similar way. Ill-nesses such as dementia, for example, lead to a loss of memory and a coherent self-narrative and therefore also a loss of attachment to home (Milligan, 2003). When people are not able to move around, home can become a disabling space (Power, 2008). Work being performed in people's own household also challenges the ideological construction of home as a space for family life, for care and comfort, and blurs the boundaries between public and private (Egdell et al., 2010). In this section, these processes of blurring the boundaries and bridging the dichotomy between public and private will be explored, both empir-ically and theoretically. It will be demonstrated that home is always, similarly, a site of meaning and contradiction, a space that is both strange and familiar (Brown, 2003; Duyvendak, 2011; Ahmed et al., 2003). Hence, I argue that the dichotomies need to be challenged and questioned in order to identify the aspects that really matter to peo-ple (Sayer, 2011), but at the same time avoiding any naturalisation or romanticisation of home (Ahmed et al., 2003). Three empirical cases will illustrate the processes of bridging and blurring the boundaries between the public and the private. Firstly, home will be discussed as a space for work. Secondly, the boundaries between working and living and between home and workplace will be questioned by focusing on the dis-courses on migrant care workers living in Austrian households. Thirdly, home will be investigated in its role as a place of resistance, as a place that is positioned as offering protection against societal pressures and processes.

Home as a work space

Blunt and Dowling (2006: 100–101) argue that public discourses present

> a dominant or ideal version of house-as-home, which typically portrays belonging and intimacy amongst members of a heterosexual nuclear family, living in a detached, owner-occupied dwelling, in a suburban location.

In the locus of home, it becomes clear that the recognition of marginalised groups (e.g. those doing the caring) is closely linked to economic inequalities. Only those who can afford it can afford the ideal of home. And these economic and social circumstances lead Parks (2002) to argue that a positive notion of home is exploitative of and alienating to those providing care work in it. Simply because care at home 'falls outside the market economy' and because it is therefore socially and politically invisible (Parks, 2002: 19), it marginalises those who have to fulfil unrecognised and unvalued work, work that is largely not even understood and recognised as work. The home has a particular ideological meaning that hides socially necessary work away from public recognition and adequate economic remuneration:

> Paid domestic work within the home not only challenges the socially accepted meanings of the home and its association with the private and the familial, but also makes plain the complex intersections of domesticity, class position and racial difference that distinguish women and create divisions between them.
>
> (McDowell, 1999: 83)

Criticism of the public/private dichotomy has a long history in feminist thinking and politics (Landes, 1998; Scott and Keates, 2004). Blunt and Dowling (2006) point to the long-existing ideological separation that men build and dwell and women preserve. The idea of home as a refuge, however, as a place of safety and comfort, is an ideological construct. Gender plays an important role in defining the various aspects that separate the public and the private (Martin-Matthews, 2007); and consequentially, the constructed meaning of home might differ for men and women (Russell, 2007). Blunt and Dowling (2006: 16) argue that the idea of the home as a retreat is a male construction, of men, 'for whom home is a refuge from work, but certainly doesn't describe the lives of women for whom home is a workplace'. The combination of home as an emotionally feminised place (Williams and Crooks, 2008) and the fact that most work within the home is done by women led to many second-wave feminists' critique of an idealised notion of home where one's own house is understood as an oppressing, exploitative workspace in which processes such as domestic violence can be experienced (Blunt and Dowling, 2006; Duyvendak, 2011). In the context of care, Martin-Matthews (2007: 246) therefore situates 'home at the nexus of the private and the public spheres', and Blunt and Dowling (2006) convincingly argue that dualistic thinking about home, which creates clear dichotomies and relations (emotions–rationality; tradition–modernity; private–public; feminine–masculine; local–global), is wrong, as both categories of this dichotomy can always occur simultaneously. Home only exists through contrast and confrontation with the outside world:

> Home is not separated from public, political worlds but is constituted through them: the domestic is created through the extra-domestic and vice versa.
>
> (Blunt and Dowling, 2006: 27)

The critique of the dualistic construction of the public and the private and its link to gendered associations is important, but itself needs to be questioned (Ceci et al., 2012). As authors such as bell hooks (1990) emphasise, criticising the notion of home as inevitably exploitative for women can be seen as a white, Western, middle-class feminist project, since for many women the home also functions as a site of resistance. On the other hand, empirical practices challenge the dichotomy. Care, it can be said, makes home public. As mentioned above, formal services fit only unsatisfactorily into the idea of the private home. Twigg (1997) describes this process as being based on spatial oppositions between public and private in the home space itself and argues that, in the process of tasks performed by care services in the private space, the home is blurred and partly loses its poignant characteristics. Practices of care can lead to the institutionalisation of the home space (Milligan, 2003), such that bedrooms, for example, become a public space (Wiles, 2003). More recent caring practices, such as telecare, very consciously bridge the boundaries between the private and the public spheres (Latimer, 2012). For both care receivers and carers, these practices change the experience of the private home fundamentally (Milligan, 2003). The home as a site of care (work) creates several paradoxes, such as between autonomy and dependency, or social policy and family relations (Brown, 2003). The blurring of public and private spaces through caring does not only cause problems for the power situation between carer and cared-for, but it can also change the meaning and experience of home in general. Similarly, the bridging of boundaries also takes place in semi-public spaces, such as care homes, where public exposure and privacy are both defining aspects of the experience of the space (Brown, 2003). The power to live a self-determined life in the context of a care home is restricted and can manifest itself in resistance to the care staff when residents try to gain control over a number of aspects of their lives and surroundings (Kontos, 1998). Phillips and Bernard (2008: 87) argue in this context that 'a blurring of the boundaries between these dichotomous spaces [...] has increasingly occurred, challenging in its wake our conceptualisations of care'. Work in the home challenges the dichotomy of public and private and reiterates that the two separate spheres are ideological and moral constructions. The ideological distinction between public and private is not therefore a spatial one but rather a discursive one. The split is not only between institution and home but also an ideological distinction, replicated in itself, called a 'fractal distinction' (Gal, 2004). Gal (2004) gives as an example of its self-replication how the private space of the house is again split into a public space (e.g. living room) and a private space (e.g. bedroom). The ideological and moral split between informal family care at home and professional and/or commodified care in the institution can be reproduced in the context of professional care itself. Some professional carers are seen as 'real' carers, others are not. Another example is care delivered by non-family carers in the home setting.

Strangers caring in 'homes'

The construction of traditional 'family care' is built around the intersection of themes of close relationships and one's own home. In this section, the consequences of strangers caring in people's own home are explored. For this endeavour, I return to the discourse on migrant domestic carers as an example of the ideal of home being created and constructed. Discussing migrant carers' roles in the household will help to understand the relationship between the place of care and its meanings and connotations (see Weicht, 2010). The analysis of the discourse on migrant carers in particular suggests that the home is constructed as a sphere in which informal care, based on affection, love and duty can be practised, even when performed by non-family carers. Bettio et al. (2006) observe a widespread aversion to institutionalisation in Italy and link this to the motivation to employ migrant carers (see also Degiuli, 2007). Similarly, analysis of the Austrian discourse suggests that home is constructed as the sphere in which informal care based on affection and love can be practised, even when performed by non-family carers (Weicht, 2010). In other words, the employment of migrants as carers in people's own home, described by Martin-Matthews (2007: 231) as 'strangers who attend to her in the most intimate settings', reproduces the idea of home as the realm of family care, and the notion of family care is therefore extended to non-family members. McDowell (1999) makes the point that precisely because the home is associated with love, emotions and empathy, the work performed within it is not seen as work but as expressing dedication and feeling. The following extract from a newspaper article reporting on the situation of migrant care workers in Austrian households clearly shows the kin-like function. Beginning with the terminology 'Granny' in the headline, the whole description of the situation suggests that living together at home is constructed as the building up of a family relationship. The last sentence then presents the alternative (here mentioned as the only alternative – a limitation often found in this particular discourse), the care home. It becomes clear that the carer prevents the cared for being 'pushed off' to a care home.

'When Anna is gone, Granny gets ill.'

Anna does really everything that comes up. She cooks, washes, does the housework. But over everything else she faithfully looks after 'Granny'. For many years Granny has been dependent on others' help. The nearly 90-year-old woman has Alzheimer's and is bedridden. Additionally, a chronic lung disease causes problems. Anna helps Granny into her wheelchair, washes her, supports her in the daily tasks. 'But especially during the night she is there', Margit says. 'That's the greatest thing for us. But only the illegals do that. Otherwise one cannot afford that.' (…) Today everything's different. 'The two are extremely close. Every three to four months, when Anna goes to see her family in Slovakia, Granny gets ill. She relapses – and that's every time.' (…) A care home, however, is out of the question for the family. Margit: 'Mother always refused to be pushed off.'

(*Kurier*, 13/08/2006)

For family members' performance of intimate, affectionate care, the most important feature is the provision of care in people's homes, and the analysis suggests that family members can fulfil their moral duty, stemming from their familial connection to the person in need, at least in part by arranging for their relatives to be cared for in their own house. Mehta and Thang (2008), focusing on the situation in Singapore, argue that, there, society approves of people's filial responsibility as long as care at home is ensured. Reflecting the Austrian discourse, *Der Standard* (07/02/2007) writes that it is beyond doubt that the goal of policymaking in the context of care is *to enable care and minding at home*. Thus, migrant carers who live with the cared-for person ensure the execution of informal care, also in replacement of family members. People whose moral duty might be perceived as being actively involved in intimate care for the elderly can be assuaged by ensuring that their loved ones are *saved from* or *prevented from* having care in an institution. And in this context, migrant carers take on the role of domestic, informal carers and are therefore able to provide the services that are usually restricted to family members. When the *Kronen Zeitung* (08/07/2007) writes therefore about *families, who sacrificially care for their relatives at home with Eastern European help*, the inclusion of migrants in the home seems to correlate with inclusion in the family. Migrant carers living *with* the cared-for person are constructed as the logical actors who ensure informal home care:

> Thank God they exist, the good women from the new EU-East (. . .), four truthfully nice supporters from Poland (. . .) lived one after the other with her and cared for her.
>
> (*Kurier*, 16/08/2006)

The constructed ideal care relationship described in Chapter 2 can therefore be established by living with the cared-for person. Due to the strong link between one's own home and real care and the importance of the former for the latter, migrant carers are constructed as the only available option for people. Anything else (especially institutional solutions) would challenge the idea of care itself:

> The first impression: There are two that really get along well – even though they see each other 24 hours a day. For two years, since Mrs P.'s stroke, the young nurse Maria cares for the 67 years old Viennese. [. . .] [L]egal 24-hour care is too expensive (and difficult to get) and she panics about going into a care home: 'I have experienced that with my mother. I don't even want to think about it.' What's left? Maria.
>
> (*Die Presse*, 14/08/2006)

In the context described above, 24-hour care is only discussed as care 'at home'. Even though a reference to care in care homes can be found, the label is almost exclusively attached to care within one's own home. The following extract is a reader's comment that appeared in an Austrian newspaper in response to the political discussion on the then illegal practice of employing migrant carers in someone's own home. It describes the possibility of people staying at home and being cared for in their own house as a situation that is honourable and which should be supported. Politicians and the political process are criticised for interfering with what is happening in one's own home. This links to the argument about the construction of public and private spheres, the idea of natural communities and the interference of politics and bureaucracy (see chapters 4 and 6). Politics is constructed as the opposite to a natural arrangement of care:

> We have really reached a point in the state of Austria! Now, apparently, you already get penalised if you don't push your helpless, old parents off into a care home, so let them be cared for by foreign care workers in their own familiar home! This falls under the sector of humanitarian help and this is, as one knows, tax-free! As many employers enrich themselves by not paying taxes the state should reduce the employment of the many thousands of illicit workers in the construction industry and other areas and not take a stab at private individuals who don't want their own flesh and blood to die dishonourably.
>
> (*Kronen Zeitung*, 19/08/2006)

It needs to be noted, though, that the honourable solution (i.e. the employment of migrant care workers to enable care at home) is not matched by a conscious acceptance of the role of employer. Rather, the desired solution of care at home is enabled by the expense of migrant care workers' reduction of rights and interests (see Weicht, 2015).

'Home' in opposition to the outside world

An important feature of the discourse on paid care work performed in particular places is the ideological relation between institutional, public spaces and the sphere of money and payment. This is important as the ideological and moral dichotomy is reproduced and re-established. The relationship between private spaces and private relations is striking, as observed by Gal (2004) who argues that private property is a feature of capitalism but that private intimate relations are 'ideally protected from economic calculation' (2004: 261). This paradox can be explained by understanding the ideological and moral creation of hostile worlds and the public 'communication in which social organisations are imagined in nested ways' (Gal, 2004: 275). Historical perspectives demonstrate that the idealisation and ideological construction of home happened in opposition to the developing capitalist economy (McDowell, 1999). While the imagination of home often shows conservative, strongly gendered attributes, the same discursive construction also provides a potential critique of modern capitalist society. I have mentioned before the potential gain of ontological security that can arise in the home, which, as Milligan (2003: 461–462) concurs, must be seen as 'a familiar and "safe space" [away] from the threats of the outside world'. Postcolonial and black feminist perspectives have emphasised that the construction of home must also be understood as empowerment and resistance by black women in the light of a racist and sexist reality (hooks, 1990). The home can thus function as an explicit space that rejects and resists infiltration by the public:

> Black women resisted by making homes where all black people could strive to be subjects, not objects, where we could be affirmed in our minds and hearts despite poverty, hardship, and deprivation, where we could restore to ourselves the dignity denied to us on the outside in the public world.
>
> (hooks, 1990: 42)

Postcolonial geography has demonstrated how home must be seen as a reflection of mainstream power relations and at the same time as a place of resisting and challenging these ideologies through everyday practices (Blunt, 2005). It is interesting, and a sign of the ambivalence associated with people's imaginations and constructions of spaces, that the traditional, bourgeois, middle-class ideal of home must then also be seen as an antipode to the capitalist world of work, employment and markets. However, Young (2005a: 156), in this context, states the following:

> Consumerism encourages people to focus on the private spheres of their homes; to this extent home is a counterpart of the capitalist marketplace and a detriment to the solidarity of community and assertive public participation.

Mallett (2004: 71) argues that the public sphere 'is associated with work and political engagements and non-kin relationships' and that home, on the other hand, is perceived as a haven against an imposing, threatening and dangerous outside world. While Mallett (2004) points out that these associations are often not reflections of reality, it is important to understand that as an image and a nostalgic feeling they have important real consequences. Young raises therefore the question of whether 'an end to such exploitation requires rejecting entirely the project of supporting identity and subjectivity embodied in the patriarchal ideology of home' (2005b: 130). In other words, does positive affirmation of the construction of home inevitably lead to gender inequalities and exploitation? In trying to identify important aspects of the meaning of home (and in light of the construction of the meaning of care), I thus want to move beyond a simple differentiation of caring or careless spaces (Conradson, 2003a). Young (2005b: 151) concludes that feminist thinking and politics should adopt a dialectical approach in relation to home:

> Feminists should criticize the nostalgic use of home that offers permanent respite from politics and conflict, and which continues to require of women that they make men and children comfortable. But at the same time, feminist politics calls for conceptualizing the positive values of home and criticizing a global society that is unable or unwilling to extend those values to everyone.

A more procedural approach is needed in order to understand home-making practices for both men and women (Blunt and Dowling, 2006) and what role care plays in this context. This approach would also mean taking seriously the dialectic of experiences of people, in particular places and their imaginations in broader society, and thus a move beyond a duality of place (that has concrete meaning for people) and space (as an abstract description) (Hanlon et al., 2007).

The discourses on migrant carers in Austria have shown the inevitability of linking a re-imagining of home to a re-imagining of close relationships. Queer geography has provided similar insights in relation to family life (Fortier, 2003). Fortier (2003: 115) argues that in order to identify inclusive ways of caring, it is necessary 'to decentre the heterosexual, familial "home" as the emblematic model of comfort, care and belonging'. Additionally, literature that explores the experiences of migration and home has provided important insights into the particular constructions of spaces of comfort and belonging (Ahmed et al., 2003). Reassessing the notion of home thus requires challenging the heterosexual, white, middle-class construction of home while not rejecting the meaning home conveys for those providing and receiving care. While everyday practices are undoubtedly important, the ideological construction of home still presents a rather static image. The ambivalence present in some accounts of public discourse needs to be taken seriously and to be understood as an attempt to link everyday practices with public moral expectations. Home's implicit connotations of relations with others can be seen as '*part of* rather than *separate from* society' (Blunt and Dowling, 2006: 14).

Conclusion

In this chapter, I have discussed the geographies of care by analysing the importance people attach to certain places in their imagination and experience of care and caring, illustrated by Williams:

> There is for virtually everyone a deep association with and consciousness of the places where we were born and grew up, where we live now, or where we have had particularly moving experiences.
>
> (Williams, 2002: 145)

Care, like all social practices, needs to be understood as physically and discursively embedded in concrete places. The places and settings in which care is delivered and received are furthermore linked to particular social relations (Hanlon et al., 2007). Thus, care cannot be understood independently of those places, and, consequentially, care also offers insights 'into the interstices of production/consumption, public/private and formal/informal that characterise places' (Hanlon et al., 2007: 479).

People continually express wanting to be cared for at home, and so a link between idealised loving, affectionate care and home is established. The discursive construction of home impacts on the meaning of care and home, which, like other places, is at least partly defined by a particular use of words and discursive practices (Augé, 1995). I have demonstrated that the meaning of one's own home is based on the intersection of physical-materialist, emotional and relational connotations. The meaning of the place of care is established by the use of value-laden and strongly gendered dichotomies, such as the distinction between the private and the public, home and institution, place and non-place. Using Marc Augé's (1995) conceptualisations, it has been shown that institutional care arrangements are constructed as quintessential places that lack intimacy and thus care. Care homes are constructed as the complete opposite to what care is associated with. The care home is seen as a manifestation of other-dependent, individualised living, whereas one's own home represents family ideals and values. This ideological dichotomy prioritises one's own home as a space of comfort and relations and, thus, care. Traditionally, these associations are linked to feminine attributes, meaning that spatial divisions between the private and the public constitute gender divisions and vice versa (McDowell, 1999). However, the discursively shaped dichotomies are continuously blurred and bridged in relation to care practices. Importantly, I have argued that the construction of a dichotomy of home and institution represents an ambivalence that is inherent to care. Feminist, postcolonial and queer literature has shown that home does not convey the same (positive) associations for everyone. Additionally, the performance of (paid) work within the home challenges the traditional connotations and their meaning. I have used the discourses on the employment of migrant care workers as an example for the renegotiation of the meaning of home and the dichotomies underlying it. It has been shown that home can create family relations, even with non-kin members. In that sense, home does not have a fixed, rigid meaning but is rather negotiated and reconstructed through concrete practices. What home means to people is affected by everyone's own historical experiences, and Blunt and Dowling (2006: 245) summarise the constitution of home as something that is *made:*

> ... home is a *process* of creating and understanding forms of dwelling and belonging. Home is lived as well as imagined. What home means and how it is materially manifest are continually created and re-created through everyday home-making practices, which are themselves tied to spatial imaginaries of home.

In that sense, not only does home influence the meaning of care, but care also plays an important role in shaping the meaning and experience of spaces (Mee, 2009; Williams, 2002; Yantzi and Rosenberg, 2008). The concept of therapeutic places has shown how some places and spaces are themselves ascribed therapeutic qualities (Conradson, 2003b). This understanding of care, also shaping the meaning of home, might enable a reconstitution of spaces of care and allow a rejection of the idea that the values of home are always linked to one's own house (Ceci et al., 2012).

Not only people's feelings and associations but also their opinions and arguments about care show an ambivalent relationship to places in general and home in particular. One's own home is often idealised in discourse as a safe haven, even though it is often not safe after all, as maltreatment and abuse can often be found there. However, the home is to some extent constructed as the safe (family) space in opposition to the cold, impersonal and cruel space of the institution. This demonstrates home's reality as a space that is continuously reconstructed as the outcome of a set of social relations linked to particular places (Egdell et al., 2010). The analysis therefore links the construction of idealised caring relations to the spatial dimensions of these caring relationships (Milligan, 2003). The process by which migrant carers are included in family settings exemplifies the importance of the construction of home for the construction of the idealised version of (informal) care. Stereotypically, home represents 'warm' feelings and positive associations with family, love and comfort. Images of being cared *for* and being cared *about* reflect similar sentiments and feelings, and home becomes a central feature of positive imaginations of care. In that sense, even 'strangers' can be incorporated in a construction of 'family care'.

By looking at the construction of the space in and through which care is provided, the point can be strengthened that family ideals, as described in Chapter 2, are also a representation of a desire for safety and intimacy in a world which is experienced as overwhelming and market driven. It could be shown that home in the context of care is constructed as a refuge from economic demands 'outside'. The home can therefore be seen as the physical and spatial expression of what care means to people, the ideal of 'being there for each other'. These values can be seen as being in opposition to the dominant, hegemonic market ideology. Success, competition and self-interest are counterpoised by a particular imagined world. This world, I argue, manifests itself in the notion of home. Similarly, Young (2005b) describes the potential of the private space to be an anti-capitalist refuge in which people can resist the enforced political and economic structures of the public sphere. She (Young 2005b: 149) argues that this resistance 'requires a space beyond the full reach of those structures, where different, more humane social relations can be lived and imagined'. In the public discourse, the focus on home as both as symbol and physical space goes far beyond an uncritical favouring of traditional family structures and ways of living (which are present nevertheless). In fact, home also bears a potential disconnection from societal marketisation and economisation. The ideological symbol of the home opposes market domination and the materialistic aspects provide an imagined shelter. Whether these values can provide a 'leverage for radical social critique' (Young, 2005b: 146) remains to be seen. In the current economic, political and social circumstances, home care can only be lived and experienced within the dominant, hegemonic social structures and ideologies. Unpaid caregiving within the home, for example, is strongly linked to the physical, material and social constraints of home (Wiles, 2003; Yantzi and Rosenberg, 2008). Ahmed and colleagues (2003: 9) therefore remind us to question 'the very *terms* of home which delimit it as an accomplished site of belonging and governance'. In other words, the emancipative potential of home, as described above, is met by the penetrating socio-economic conditions. Parks (2002: 28), for example, points out that 'the high rate of at-home care by black family members may be a labor of love – and may be wrapped up within an ethic of family and community', but it has serious consequences for those disadvantaged and marginalised in the first place. Generally, the unifying values of home inevitably create an 'other' that is excluded from this 'better world'. At present, a positive reference to the home without some affirmation of traditional excluding and exploitative conditions seems unlikely, if not impossible. Under the conditions of a neo-liberal capitalist economy, the construction of home, as discussed in this chapter, reinforces divisions and stratifications and reproduces social inequalities in terms of gender, class, ethnicity, disability and age. In particular, for women, the ambivalence of both home and care at home needs to be borne in mind, and any policy intervention needs to start with the recognition of these ambivalent associations with home.

Because home represents a particular ideal of real care in association with values such as family, community, independence and intimacy, policy thinking and policymaking need to be conscious of the implications of care at home. The discursive construction is powerful and is shaping people's ideas, imagination and experiences about care. At the same time, the 'ideology of home' is also, as Robertson (1995) argues, a response to recent claims of 'homelessness' and 'rootlessness'. Duyvendak (2011) raises the question of whether home is just a discursive trope reflecting a nostalgic imagination and longing. Home represents an image that is both nostalgic (as it might represent traditional family ideals) and progressive (in opposition to a neo-liberal world). Care practices and the (social and political) organisation of care are influenced by both, but also draw on both aspects. The next step will be a further exploration of these feelings about nostalgic imaginations of the ideal of care. An extension of both family relations and the ideal of home can be seen in the notion of 'community'.

4
How Should Communities Care?
Nostalgia and Longing for the Ideal

Introduction

Often when I tell people about my research, they emphasise the importance of the topic of elderly care in the years to come. Care, it is argued, will prove to be an enormous challenge for societies, due to ageing societies and changing family and social structures. Potential reactions and solutions such as public arrangements of care are generally seen as second best option – dissatisfying in some, deeply personal sense. In particular, the public and professional provision of care is often compared to how care *used to be*. A picture of earlier generations is sketched out, when people lived in families and communities that still cared for their elderly members. Drawing on nostalgic constructions, communities are imagined to be a safe and comfortable place, a place where people are there for each other and where care can actually take place. At the same time, however, people do not envisage a return to past times; rather, community in its emotional construction is something gone, something to be longed for but at the same time firmly situated in the past. With respect to relationships in care discourses, I argued earlier that 'family' does not necessarily refer to *who* is providing care but to *how* care is provided, that is, what constitutes the ideals of care. Similarly, the discussion of the geographies of care has shown the importance of the home as an idealised space in which real care is possible. In this chapter, I revisit these themes of imagination and ideas and, moving beyond the immediate family, I will be turning to the notion of community, more precisely to an expression of a societal yearning for community. In that sense, the idea of community can be understood as a combination of features linked to family and home. Comfort, closeness, shared responsibility and concrete relations establish and give meaning to the idea of community.

Community has been an important concept in relation to care for many decades (for classical descriptions of the importance of community, see Townsend, 1955, 1962). In the United Kingdom, as in other European countries, for example, 'care in the community' has been an ideal for policy provision in relation to older people's needs (Baldwin, 2000; Bauld et al., 2000; Bornat et al., 1997; Parker, 1999; Springings and Allen, 2005; Stackhouse, 1998). Care in the community has been widely promoted in the political arena as it promises to be a cheaper option (for public funds) than professional and institutional care. An emphasis of values and virtues related to community has therefore recurrently been utilised to justify cutting down public services (see Heaton, 1999). At the same time, support for care in the community relies on imaginations and idealisations that closely resemble desires and needs related to care itself. In these contexts, community describes an attitude of people who share specific spaces, such as neighbourhoods, people who are there for each other. The terminology of community is also widely used in people's conceptions of care, in imagining an ideal society and in people's references to a better social life. Often a description of community sketches a nostalgic idealisation of past times or other places, related to moral disapproval of current welfare practices and social interventions. While the latter are often seen as necessary but still loathed bureaucratic solutions, the associations conveyed in terms of family and community satisfy the desire for ideal care. In that sense, community is contrasted with modern society, characterised by individualisation and globalisation. Individualism, for example, is, in that sense, seen as a product of economic modernity and related changing family structures. Giddens (1998: 36), for example, shows that

> [t]he new individualism, in short, is associated with the retreat of tradition and custom from our lives [...] The welfare state has played its part: set up under the aegis of collectivism, welfare institutions have helped liberate individuals from some of the fixities of the past.

The trend towards increasingly organising and arranging social policies and social interventions on a local rather than a national level takes up the discursive meaning that community and related associations convey. As in the case of home, community is a physical entity, often a particular living arrangement within a specific area, such as a neighbourhood. Community, however, also refers to a conception and agglomeration of particular values, feelings, emotions and associations (Duyvendak, 2011). In a recent book, Arlie Hochschild (2012) describes fundamental changes to social cohesion and communal living. Relationships are described as being more fluid, people are (both physically and virtually) increasingly mobile, markets penetrate more and more areas of people's lives and social and emotional practices that were formerly thought of as expressions of personal relationships become commodified and thus can be bought and sold in the market. Importantly, however, all of Hochschild's examples are still compared to and judged against earlier, familial arrangements, both real and imagined. Nostalgic sketches of the past function to describe and conceptualise current wanted and unwanted social, political and cultural developments.

In this chapter, I discuss the narratives, emotions and values that constitute the positive feeling that community provides and their significance for the construction of care. The chapter starts by identifying the discursive patterns, narratives and images that emphasise the significance of community within the discourse. What does a longing for more community living mean? What are the consequences of imagining better times in which people in communities care for and about each other? Secondly, turning to the dichotomies underlying the construction of the meaning of community, the concept of nostalgia is used to describe processes in which community is related to feelings of societal decline. A strong focus on public and political discourses embraces the goal to 'regain [the] community we've lost' (Sprigings and Allen, 2005). The feeling and subjective experience of a decline of community is closely linked to the cultural, social and ideological meanings of care and its discursive construction. What role does a nostalgic focus on temporal distance (the past) or geographical distance (other countries) have for the meaning of care? Thirdly, the dichotomies constituting the longing for community and neighbourhood are blurred, both theoretically and empirically. Community is not inevitably linked to positive, comforting experiences. Additionally, in the discourses around care and carers, community is always something that is gone; it is an ideal that a society should strive towards but can never, due to its idealised conception, attain. Finally, the longing for community can also be described as the imagining of another life, a life that is desirable and a life in which being there for each other, that is, caring for each other, is possible. I will investigate how far a 'productive nostalgia' can be identified, an imagined place in which community functions not only as an idealised conception of the past but also as a promise and an ideal for the future. In that sense, I will discuss which aspects of the nostalgic imagination of community are crucial for a re-conceptualisation and re-organisation of care in current times.

The meaning of community

The meaning of community is closely linked to desires and imaginations about the perfect way of societal living. A community is constructed as a realm in which compassion, support and mutual affection dominate this mode of living. The positive and embracing notion of community relates to imagining a place of comfort, free from the pressures associated with social, economic and political processes of commodification and marketisation, and thus references to community are often linked to negative sketches of modernisation and capitalist development. In his highly influential conceptual distinction between *Gemeinschaft* and *Gesellschaft*, Tönnies (1955) describes the ideological separation of a community based on personal, direct social relations and shared values and beliefs (*Gemeinschaft*) from an association based on indirect, impersonal ties and rationally constituted rules (*Gesellschaft*). In this dichotomy, communities are constituted through affection and feelings and represent the non-modern world, whereas the modern conception of society is based on rationality and agreement. Bauman (2001: 1–2) takes up the concept of community, describing its strength in creating safety in a dangerous world by making it possible that 'we are never strangers to each other'. Modern political, economic and social developments are seen as markers of a culture that has alienated people from each other and their community units. Economic and social demands linked to an ethic of individualism are positioned as the antipode to earlier forms of reliance on family and community (Beck and Beck-Gernsheim, 2001). Several discursive aspects comprise the imagination and construction of community as a manifestation of non-commodification. Community represents more than just a *Gemeinschaft* of individuals who share some features with each other. Rather, it reflects a moral ideal and an image of a good society which is positioned outside the market and based on an idea of general altruism and taking responsibility for each other (Firth, 2007; Jones, 2008; Nisbet, 1966). The reference to a longed-for community is based on a desire for particular values and virtues and an imagined better life. Bauman (2001: 1) argues that

> ... '[c]ommunity' feels good because of the meanings the word 'community' conveys – all of them promising pleasures, and more often than not the kinds of pleasures we would like to experience but seem to miss.

Both in politics and among the general public, community is a recurrent concept in explanations of the best ways to live in old age and to organise care. Ivan Lewis, from 2006 to 2008 the minister for care services in the United Kingdom, makes this link explicit in a commentary published in the newspaper *The Observer*:

> ...there are few more important challenges than the way society treats older people. (...) [C]ommunity networks, led by the voluntary sector and faith groups, should be supported to deploy volunteers and 'good neighbours' to tackle loneliness and social isolation. It is not the state's job to provide befrienders, but it is the duty of any community that has a right to the description 'civilised'. (...) [W]e want older people to be valued as active citizens, mentoring and acting as role models to young people and, likewise, young people to be supported to befriend and 'adopt' older people.
>
> (*The Observer*, 24/06/2007)

In this example, several values and virtues associated with community can be identified: the neighbourhood is constructed as a realm in which community happens; faith groups and (unpaid) volunteers are, apart from family members, the main actors in securing the community. The last sentence of the extract also emphasises the importance of a notion of intergenerational relating. A close and well-working community is seen in this example as an essential asset of a civilised society.

In the following sections, I will identify the main pillars that represent the foundations of the discursive construction of the meaning of community, starting with an image of the extended family. Secondly, community stands for a particular caring environment, and thirdly, the idea of community refers to a notion of natural growth and coherence, an entity which has developed organically.

Extended family

Ideal arrangements and provisions of care are often imagined to be echoing family and family values (Cowen, 1999), and as such Roos et al. (2006) argue that family also remains central to a sense of community and civic society. Since both family and community entail feelings of mutual obligations and a sense of belonging and commitment (Malone and Dooley, 2006), community, as a discursive framework encapsulating family values and virtues, is constructed as an extended family arrangement. Tönnies (1955) links the family to a notion of *Gemeinschaft*, that is, values associated with the idea of community:

> Family life is the general basis of life in the Gemeinschaft. The village community and the town themselves can be considered as large families.
>
> (Tönnies, 1955: 267)

The link between community and family is twofold: not only is community seen as an extended broader family construction, but it is also constituted as a conglomeration of different family units. This already points to a rather organic, naturalist construction of the ideal of community. Discursively, a community is created in which people mean something to each other, show responsibility for each other and care for each other. To give just one example from the *Daily Mail* referring to the treatment of elderly people in the community:

> How can we say we are civilized when we treat our elderly no better than prisoners?
>
> (*Daily Mail*, 23/01/2007)

The use of personal pronouns, as shown in the example, is instrumental in creating an imagined community encompassing all those involved in and addressed by this commentary. It emphasises the significance of familial bonds and bonds beyond the family in the context of care. The treatment of *our elderly* people is *our* responsibility as a community. Mistreatment and neglect reflect on *us*. Those who do not treat them in a particular way are implicitly described as uncivilised. Community can thus be understood as an extension of family values, going beyond the family, describing an idea of 'being there for each other' within a particular neighbourhood and setting. In an early conceptualisation of this link, Nisbet describes that community entails

> all forms of relationship which are characterized by a high degree of personal intimacy, emotional depth, moral commitment, social cohesion, and continuity in time. Community is founded on man [sic!] conceived in his wholeness rather than in one or another of the roles, taken separately, that he may hold in a social order.
>
> (Nisbet, 1966: 47)

Pessimistic accounts about current social life favour a notion of community that could be found in the past, as opposed to the insecurity and uncertainty of the present. The idea that close family ties gave security and safety is largely an ideological construction; however, as with all myths, these ideas bear an important ideological and moral function in dealing with the demands of modern society. In other words, even though these ideas can be regarded as myths, they offer people the possibility of imagining other options and thus help to reinstate a particular morality. In the following extract from a focus group, Nathan's comment clearly shows these discursive links between family and community and the imagining of the past. His description of community combines the family, the neighbourhood and local welfare arrangements as examples of past ideals:

> Nathan: And the reality is that, whereas though I wasn't around 70 years ago, the family units tended to care and [...] the people on the street would actually care [too]. In fact, my first [job], was in an environment where families were close-knit and if someone was ill [...] the local individual's family and the neighbours would actually pop in, they would do the cooking, they would bring meals, they would clean. The nursing staff would pop in, the district nurses. And everything seemed a lot less complicated. As before, the European Parliament and everything changed, brought in health and safety and cost factors [...]. But a lot of it used to be handled by family and I think, to a certain extent, I don't have a big family, but what I would like is that kind of personal care.

Nathan's nostalgic description of past times points to a dichotomous construction of community then and bureaucratic arrangements now. While I will discuss the importance of this discursively constructed dichotomy below, it is quite clear already at this point that family and community are associated with concrete belonging and concrete relations between people such that one cares about and for the other. Nathan's reference to politics and regulations contrasts the world of bureaucratic measures with a feeling, an emotional recognition of closeness. Bauman (1993: 151) also describes this notion,

> which represents community as a unit held together by the *awareness* of unity, by a fraternal sentiment which makes it family-like without making it a family, as a territory of unqualified cooperation and mutual help.

Dench et al. (2006) in their study of developments in the East End of London emphasise the significance of community, neighbourhood ties and family relations for the experience and feeling of safety and security. Community, they argue, is modelled after the image of the family, and community and community relations therefore attempt to balance the demands of modern capitalist society:

> Family ties gave people the support and security which made life tolerable, and provided a model for organizing relationships with close

neighbours. Being a member of a family gave you kin and quasi-kin locally, and made the world a safe place.

(Dench et al., 2006: 103)

The ideological and moral meaning of community relies heavily on an idealisation of past community as extended family. Whether or not the ideal created post-hoc actually fits the reality is not the principal question; rather, the ideal of community derives its significance from a discursive consensus that other times (or other places) managed to be more caring than today.

A caring environment

Mirroring de Certeau et al.'s (1998) work and their focus on every-day life where they identify the neighbourhood as an extension of the home, the logics of the private, caring space can be captured by the con-struction of community. They argue that community is a secure, safe and restful space, a space in which people seek refuge and care. Simi-lar to an aspect of the construction of home which I discussed earlier, one feature of community depicted as a caring environment is a refer-ence to the countryside. The (village) neighbourhood, the quintessential realm in which the ideal community can strive, is often contrasted with anonymous city life. This distinction is also present in Tönnies's (1955) discussion of the association of *Gemeinschaft* with rural villages and *Gesellschaft* with emerging cities. The discursive associations are repro-duced in the sense that the rural, as the ideal place for community, is associated with neighbourhood, care and families, as exemplified in the following extract from a focus group discussion:

Walter: And, I must say, caring at home, if possible somehow, that the person is allowed to live by himself, where he is visited, once in the morning and then in the evening the son comes by, or someone else. That is definitely the very best possible [...]
Vanessa: I believe that this can actually work very well here in the countryside [...]
Barbara: True. Very true. [...]
Vanessa: There is still neighbourhood, there are still families that do mind [...] and take care. [...] In the city this is of course extremely different.

A clear dichotomy between the rural and the urban is constructed where the latter is associated with, as Raymond Williams (1973: 291) put it, capitalism or bureaucracy or centralised power, while 'the country' [...] has at times meant everything from independence to deprivation, and from the powers of an active imagination to a form of release from consciousness.

From the focus group extract above, it can be seen that the construction of community mainly works through certain associations. Vanessa mentions neighbourhood and family as aspects that do *work* in rural areas but which, almost by definition, do not exist in the city. This association could be related to an understanding of the rural community as the natural, in contrast to the human-made city, in which negotiation, politics and economic participation dominate. Lefebvre (2000: 190) speaks of the city's independence from natural cycles and argues that the concept of urban society is based on an imagined evolutionary or historical development, stating that '[u]rban society rises from the ashes of rural society and the traditional city' (2000: 189). The contrast between country and city, as Williams (1973: 289) argues, 'is one of the major forms in which we become conscious of a central part of our experience and of the crisis of our society'. Jones (2008: 34), similarly discussing the 'imagined traditional community', which is always thought to be threatened and in demise, emphasises that the construction of rurality reinforces and strengthens the symbolic power of community. The strong discursive links between the rural, the community and a notion of people being there for each other remain very persuasive: 'The idea of community seemed to represent taking responsibility for other members, a sense of belonging and familiarity' (Jones, 2008: 30). Many of these associations are combined with an identification of the community with the natural.

Also relevant in this context is criticism of the disappearance of the communal infrastructure, such as local pubs, shops and post offices. The (rural) community is identified with safety, relations between people who know each other and people who share something with each other. But as de Certeau et al. (1998: 142) also emphasise, these notions are felt not to have a place in modern life anymore. As already pointed out above in Ivan Lewis's commentary, the decent society is discussed as a tight unit that has to and wants to look after its elderly people, identified as people who share some (family) relations with the rest of the community. In other words, the decent society is built on an understanding of care for 'our elderly'. The relevance of community as a particularly safe, secure and comfortable neighbourhood is also discussed in the following focus group extract, in which John explicitly refers to 'close-knit communities' that have ceased to exist. Particularly through Nathan's reply, it also becomes clear what security in the context of communities and neighbourhoods means to people. It describes an assurance that people care for each other and are there for each other.

> John: There was an instance I know in our [community] [...] and Mum lived on her own, she had no central heating, it was just a normal coal fire, and the actual neighbours, bless them, were actually going in and giving her at least one meal a day and they, that's the old-fashioned way of how it used to be [...] on a larger scale before. I mean they used to say you can leave your back door unlocked and people would just walk in, but now you can't do that and I suppose [...] from the 1960s, when they started to build these high-rise flats and started to flatten the slums and everything, then suddenly the close-knit community was just sort of scattered. And so in a sense, we've lost a lot of that, of that close-knit community, but here there is a strong sense of community within the warden-aided places. But out from that, I would say, uh, there isn't that.
>
> Nathan: The warden-aided places are great here because they give security and they give security not only in the sense of the thinking but there is a knock on the door every morning, there's conversation, there's community.

Interestingly, here a rather positive description of old-age living outside one's own house can be found. 'Warden-aided' accommodation is discussed in the context of care, and it is clear that the positive attributes associated with this form of living closely resemble the values and characteristics of traditional home and community. Security and a sense of belonging can be substantiated within established communities that have developed over time and in which people can draw on personal, affectionate ties. This means, however, that a crucial feature of functioning, organically working communities is seen in the natural growth and development of these forms of living and relating. Conversely, community's association with nature and natural ties rejects individualised, rational living arrangements, such as those found in cities.

Naturally grown communities

Community as an extension of family values and the rural neighbourhood as the ideal realm for community are specific discursive manifestations of narratives related to feelings, wishes and hopes about what constitutes community. Tönnies' (1955: 17) work is again useful here, as he argues that *Gesellschaft* as 'formed and fundamentally conditioned by rational will' can be positioned in contrast to imagined natural communities, which he calls *Gemeinschaft* 'in which natural will predominates'. This separation of the bureaucratised constructions of late capitalist societies from so-called naturally grown communities resonates with Thompson's (1991) story of the struggles in traditional working-class communities against capitalist developments. He describes these communities as 'defending their own modes of work and leisure, and forming their own rituals, their own satisfactions and view of life' (Thompson, 1991: 85). This narrative is strongly linked to an intergenerational idea of a 'natural' community in which *all* come together. Communities are thought of as spaces that have grown, in which people relate to each other as social beings through common histories, ways of life and memories (de Certeau et al., 1998). I have already pointed to the notion of the 'natural' a few times (especially in the discussion on relationships). It is important to understand that people continuously refer to a notion of naturalness which touches on biological categories (e.g. kin relations are mentioned) but which extends beyond the merely biological. People 'naturally coming together' and caring for each other is contrasted with negotiations, contracts and bureaucratic regulations. The narrative of natural communities moves beyond the nuclear family and bridges generations and other separations constructed through modern life arrangements. It expresses a feeling of being together regardless of social identities and attributes. The following discussion between Mary and Marion is an example of the discursive realisation of this notion and its links to (naturally) grown communities:

> Mary: I actually don't think that it works so much in one direction. I think, at least in Austria, it's pretty much split, that it is either very much outsourced to the family, in rural areas for example, or in the city for example, that it works very much via institutional care. And I think if there are any compromises, then only bad ones. That's my opinion. And I think that basically, until now, there are no possibilities to somehow combine it more with each other. To have good care in living arrangements in which [...] old people are integrated, with professional supervision, just like all, or most of the people would wish for. I think there is no middle course at the moment.
>
> Marion: Whereas in my village, I also rather grew up in the countryside [...] a house was built, where now the old people live [...] from around there. They weren't really uprooted, they are still in the same village, they just have a new apartment now, they live together

in this house, these are 10 people, and the families got together, and they are always looked after, and yes, they have a timetable, who has time and when, and then they come and help. Yeah, they have there, 10 families, have found each other, that was then built by the council and I believe that is not so bad [...], they can at least live alone like that.

Clear distinctions can be noted between institutional arrangements and family care, and the impossibility or at least difficulty of combining these aspects. Marion then challenges this dichotomy referring to her experiences in the village she comes from. Interestingly, the example she describes shows significant aspects associated with ideals of family, home and natural community. When she argues that elderly people are not *uprooted*, this association with naturalness becomes obvious. The idea that everyone helps each other and everyone is there for each other is at the core of the construction of community in particular and the construction of caring in general. In another discussion group, Helma emphasises the relevance of natural growth for communities. Shared accommodation, as in this example, can therefore, under certain conditions and circumstances, become a community.

> Monica: I mean generally a shared accommodation [...] if it works like that
> Helma: if it **has grown** like that
> Monica: that's great!

Dichotomies: Nostalgic communities vs. modern individualism

Sprigings and Allen (2005) demonstrate that most political projects that build on the idea of community have taken it for granted that community is something good. In the political discourses, community is often positioned against modernisation and globalisation, and it is often 'constructed defensively in relation to outside threat' (Sprigings and Allen, 2005: 407). The positive associations with community are often notions that are formulated as a reminder of the past – as de Certeau et al. (1998) emphasise, these notions are felt not to have a place in modern, individualised societies anymore. Both the traditional family and the traditional community are felt to be under threat from economic developments (e.g. globalisation, see Robertson, 1995) or are felt as having existed in other times or in other places. Dench et al. (2006: 4) in their discussion of changes to a particular London neighbourhood also highlight the ideological link between modern 'impersonal welfare provisions' and the loss of traditional 'kinship support'. Whereas Barrera (2008) positions the church and theological ethics as a moral counterweight to individualisation and the market, Dench et al. propose a rediscovery of 'small groups as a source of civic virtues':

> The culture of individual rights has obscured the value of family ties and local community for many people. The most practical way to resist that culture may lie in strengthening family. It is significant that one of our strongest findings is of the value of family and community ties in keeping ordinary people in control of their lives.
>
> (Dench et al., 2006: 232)

The opposition of family and community on the one hand and individualism and modern economy on the other is crucial to the construction of the meaning of community. Community can be seen as a nostalgic imagination of the good life, a myth that 'bring[s people] together and reinforce[s] social solidarity' (Coontz, 1992: 6). Uneasy and critical perceptions of modern times are contrasted with another situation, often located in other times or places. Public interference, for instance through political, economic or legal regulations, challenges the notion of both home and community as caring places. In this section, I discuss the dichotomous construction of community in which current times are contrasted with a nostalgic romanticisation of the past. Nostalgia relates the imagination of community to a disapproval of current arrangements. Importantly, nostalgia can be understood as being both an individual psychological disposition (Pourtova, 2013; Zhou et al., 2012) and a social configuration. Davis (1979) convincingly shows that nostalgia is not only a psychological, personal expression but also needs to be understood as a social emotion as well – a widely shared attitude in society (Bookman, 2008; Mand, 2006). In organisations, for example, nostalgia is often the common denominator for resisting change (McDonald et al., 2006). The imagination of an ideal (often a utopia) is vital for the constitution and construction of group life and attempts to work on a more related environment (Brenner and Haaken, 2000). Longing for an idealised situation can include temporal and spatial dimensions. In that sense, the expression of nostalgia can be understood as a social imagination of temporal distance and displacement (Boym, 2001). The temporal dimension as a yearning for the past can relate to idealisations of concrete periods (Bartmanski, 2011; Todorova and Gille, 2010); more often, however, nostalgia relates to an undefined period, where the past acts as a manifestation of social ideals, values and virtues.

In discourse, community and the impossibility of ideal care are often related to a feeling of discomfort with developments in society. In the following text, I want to cite an extract from a discussion group in Austria at length, as this demonstrates the many associations between the discontent with the situation of care for the elderly and general societal developments. This discussion starts with Claire talking about an experience in her work place (a care home), which relates to the theme of intergenerational 'natural' community, as mentioned above. Alfred mentions the role animals can play and later associates this way of living with rural communities. Interestingly, Claire blames the development of local economies for the changes in societal structures and developments:

Claire: Recently I had a really lovely experience with a two-year-old child. [...] One of the residents was visited by her grandchild, with the great-grandchild. And she is looking for granny, and I say, they all sit outside today and you can of course have a chair and you can take your daughter with you outside, I say, nothing can happen. And yes, they were sitting outside for two and a half hours, which had never been possible, and the mentioned lady, who screams after five minutes, please nurse, in the bed, please [...], she's sat there, and watched the child placidly, but how, only from the facial expression, from the gestures, how satisfied, how happy she appeared. [...]

Alfred: yes, but that's the same with animals

Claire: Yes, we do have dogs and cats every once in a while.

Britta: That does also work the other way. You can't learn more from anybody than from the elderly people, they have just experienced so much already, and they can communicate so much [...] and give you something, that is so valuable.

Claire: Yes.

Britta: I do think, there are really great projects. Where care homes and nurseries are next to each other. [...] And I think that it, in times like these, it is difficult to leave it to the family alone [...]

Alfred: Yes, and I think that it rather works in rural communities, where people know each other, where the groups are, so to say, small and where people indeed, one generation after the other, grow up.

And do know each other, not like in the city, in the city area, where people are more or less anonymous there.

Claire: yeah, it is in the countryside, because of the economic structures, I can only talk for us, Alfred, everything's changed a lot, because look, we used to have the industries in town, [...] You had factories, you had everything, even if the woman went to work, she was in the town. You still had the corner shop so you could quickly send the granny to go shopping. With a list and money in her hand, you sent her shopping and the children were still in town. Now with centralisation, people go, [...] the women drive to Vienna to [supermarket chain] at 5 in the morning on the first bus, come home on the last bus, the old person must be sent away to a day centre or a home, the children must be sent away to the kindergarten. [...] Whatever you want to call it, but they don't have a shop where I live now, they don't have a pub, a centre of communication, because there I also could send a child that didn't go to school yet to go shopping. I would say be careful, you have to stay on the pavement there. Or the old granny has just taken the child. That was still possible. But today, to send someone into the centre, that's already dangerous.

Alfred: Yes, yes, it also used to be [...] that you didn't need to stay on the pavement [...] because, when I was I child, we also played in the streets, also in the main street.

A discursive pattern drawing on dichotomies is established in which current societal and economic demands are contrasted with former times of care, family and community. Institutional childcare and institutional elderly care are discursively positioned in opposition to the natural, rural community. The reference to changes in the economic situation, in the discussion above, also reconstructs and reproduces the dichotomy of capitalist production and ideal community. Similar to the temporal dimension, the spatial dimension of nostalgia refers to concrete meaningful places, which, drawing again on Marc Augé (1995), could be described as anthropological places. The reference to particular places can again be focused on a very specific space, for example a particular market (Watson and Wells, 2005), or can manifest itself more broadly as an imagination, often expressed as a defence of the local against the global, as mentioned earlier (see also Massey, 2007). Distinctions between rural and urban play an important role in this dimension of nostalgia. Similarly, Caroline, in the following extract from another discussion group, refers to the economic situation and compares the demands in Austria with the advantages of 'having less' in a particular African country she has lived in:

> Caroline: Because it's simply like that, the richer we are, and the better we have it, the less we can care for the elderly. Because in [African country], the family doesn't have money, they live in tiny houses, [...]. I just have to say that the expectations are just different, because the people also don't want huge houses, down there. You just have the time, firstly, they also don't have to clean their small houses the whole day, they're done in an hour, they simply have a very different way of life. [...] Because the more you want, the bigger everything has to be, the more work it is, and the less time you have for your family. [...]
>
> Gita: it's also that you not only don't have time for care, but also not for the children, I mean, [...]
>
> Max: yes, but then you also must say that 100 years ago it was like that here as well.
>
> Caroline: It also has to do with what we are used to [...]
>
> Max: Yeah, it's not that long ago that it was like that here as well [...] the extended family [...]
>
> Brenda: yeah, in the countryside, in the countryside definitely. In the city that's a long time gone, but in the country there were extended families.

The imagination of community must be seen as a longing that is established in contrast to people's materialistic life situations. The present life conditions are 'felt to be, and often *reasoned* to be as well, more bleak, grim, wretched, ugly, derivational, unfulfilling, frightening, and so forth' (Davis, 1979: 15). As Alleyne (2002) points out, descriptions about the construction of communities are always displaced in time (past) and space (other countries). Nostalgic accounts are obviously not a literal reflection of reality. Remarkably, the construction of the ideal community operates by sketching a certain image of the past and linking changes to economic and social developments. In the context of elderly care, it is clear that people have always been dependent on family, local social networks and neighbourhoods to get support to live their lives. The notion of a longing for community is then often related to times past when people could rely on these existing communal networks. The following newspaper extract provides an interesting example of how aspects of temporal nostalgia (longing for past times) and spatial nostalgia (situating the desired state in other countries or cultures) intersect. Both aspects, however, refer to the imagining of a situation in which people care(d) for each other:

> One of the yardsticks of **a civilised society** is the way that it looks after its elderly. **A decent country** would ensure that its old and infirm received the best possible care, not least as a mark of respect that should be afforded to the elders of the community. Judged by this standard, Britain is becoming progressively less civilised. For British citizens, the experience of ageing is increasingly beset by hardship and neglect, both at the level of individual families and the institutions of the state. **In other European countries or in Asian societies** where family life is still very important, people venerate their elders and assume it is their duty to look after them when they can no longer look after themselves. In Britain, by contrast, expectations have changed along with a profoundly altered way of life. **People are too busy and too self-centred** to assume such responsibilities. In particular, many women who once would have assumed it was their duty to look after aged parents are now themselves in paid employment. In addition, **family breakdown** is increasingly snapping the vital bonds of attachment between generations. [...] As the Health Service staggers under its own financial crisis, elderly or chronically sick people are being discharged from hospital into 'community care', only to find that the community doesn't care at all and that neither nursing nor other essential services are available.
>
> (*Daily Mail*, 11/01/2007, my emphasis)

The notion of 'civilisation' is particularly interesting in this context as it often refers to an idea of progress and progression of societies and countries. In relation to care, civilisation is often linked to an ideal of real care, situated in the past, whereas current socio-economic developments have pushed society away from civilisation. This temporal change in the references to civilisation functions as a means to establish and reproduce the dichotomy of community, care and family on the one hand and modern society with all its associations on the other. The ideas of *civilised society* and *decent country* are linked to a particular arrangement for elderly people. A historical perspective is applied, which tries to show that the change in lifestyles in the United Kingdom has led to a situation in which community does not care anymore for its elderly members. At the same time, other countries and cultures are constructed as desired havens for elderly people. These (inevitably rather abstract) places are described in opposition to modern Britain, as being based on a culture that not only deals differently with elderly people but also shows a different public morality in general. The link between imagined countries and the past figures prominently in debates on arrangements for elderly care, following Alleyne's (2002: 611) observation that people feel that 'we' have individuals while others have community (which we once had). Finally, the provision of care is linked to a broader discussion of social conditions. Self-centredness, economic involvement and family breakdown are linked to an image of a busy, self-absorbed and selfish modern society. The imagination of a community that is gone but that is still to be found in other parts of the world expresses, on the one hand, a longing for a particular state of community and society and, on the other, recognises that this desirable state of society has been lost. In a focus group discussion, Caroline again refers to the example of an African country to sketch an idealisation of living together and caring for each other:

> Caroline: Down there, firstly, unemployment is different there, secondly, there's a different living situation there. It is very normal down there for example that you do have a yard, where, five entrances come together, from different houses. The houses are of course considerably smaller than here [...] and in one lives the aunt, in the next one granny, there the sister lives and there the brother lives, so the whole family lives there. [...] And all of them on this yard together. It also means that, as far as children are concerned, it doesn't matter at all, whether it's about an old person or children, the care, the social willingness to be there for each other, is very different. Because they have the opportunity, though.

The link between economic development and work demands and the possibility of living together and for each other (i.e. community) becomes obvious in Caroline's comments. Spatial nostalgia is situating the ideal community in other countries and/or other cultures. It is emphasised that this (better) way of living, this communal lifestyle, is nowadays impossible in people's own societies. A particularly striking feature of the discursive pattern of situating the ideal community in other countries is that the ideal that is long gone in people's own society can still be found with reference to different 'cultural' groups in 'Western' societies. Larry in the following extract mentions different ethnic groups as having different constructions of community:

> Larry: It varies, across the board, through different [...] levels of society, and also, different ethnic groups as well. [...] If you look at the

Indian community you'll probably find [...] generations, all living in the same house.

Pamela: Oh yes.

Larry: But there, there they have this ethos, ethos of care, going all [...] the way through I think. Same for the Chinese as well.

Nostalgia constructs a dichotomy between the desired state of community and current economic, social and interpersonal relations. In practice, however, the dichotomous separation does not hold. Rather, the moral and ideological hierarchy associating community with an ideal state of existence is blurred; notions and practices of community are constantly negotiated and desires and imaginations are redrawn.

Beyond the dichotomy

I have so far argued that the discursively constructed dichotomy places an idealised treatment of elderly people within the community, in opposition to the selfish, rationalistic, economic world of employment, labour and busy living. The longing for community constitutes morality and 'reinforces moral commitments and inclinations' (Smart, 1999: 168) and could lead to a new moral framework. The meaning of community is fundamentally based on a commitment towards each other (Bauman, 2001) and a duty to help each other, while political and economic competition is constructed as belonging to a sphere of rationalist, materialist decision-making. However, in people's experiences, these dichotomies are necessarily bridged and challenged. Rejecting community and embracing individualisation and being anonymous also offer potential advantages to some. People might desire not to be part of the community or at least not to belong to a community all the time:

> How do we distinguish between idealized conceptions of community that sustain democratic work, *vs* those that disguise domination through rousing appeals to group harmony and community?
>
> (Brenner and Haaken, 2000: 335)

Similarly to Bauman's (2004: 62) claim that '[f]or most of us [...] 'community' is a Janus-faced, utterly ambiguous phenomenon, loved or hated, loved *and* hated, attractive or repelling, attractive *and* repelling', in the public discourse the disadvantages of community are discussed. Bauman (1993, see also Smart, 1999), in his discussion of ethics, describes the role of community as having changed, from an individual's security to an individual's burden. The ambiguity, I would argue, also means that naively embracing community is not possible either. Restricting one's independent autonomy, community also imposes pressures and demands on individuals. In an example from a focus group discussion, Claire, herself a nurse in a care home, emphasises her own need for respite, which would be challenged if she was employed in a care home in her own community:

I: And does the distance also play a role, from the people one cares for?

Claire: Indeed, I did question that for myself, because I might have had the possibility to work in [her home town], we do also have 2 care homes. [. . .]. And I had worked for 15 years in the other village, and I thought no, I am glad that I'm away from home. Because indeed, it is like that in [work place], I am at work for 11 hours there, I do that for myself, alright, but when I leave, I leave the baggage inside. [. . .] And when I now go to the farmers' market in [work place], and meet some colleague or something, that's ok, but I can't meet any relatives, who can complain. [. . .] In [home town] that would have been very different. At every corner, 'this doesn't work', and 'that doesn't work'. I've seen that, I've worked there for one month on trial, only to see for myself, whether I want to work in elderly care at all. And even this one month I have experienced it, the aunt comes, this person comes, everyone, and I think, no, that's not what I want. There, I'm anonymous. There I do the care work, there I'm known as nurse Claire, that's it. When I come out, I can go cycling, can call [a relative], can go to Vienna, it doesn't matter.

Claire points to some aspects of community, which she experiences as negative and challenging to herself as a person. I argue that this stipulation, which is based on her position as a care *worker*, emphasises and reproduces the dichotomy between community as the realm of personal relations and the bureaucratic, individualised realm of work and employment. Like all other discursive concepts, community gains meaning (both positive and negative) within specific circumstances and within particular, concrete relations. While community as a progressive, modern counterforce against economisation is seen to provide security and cohesion within a risk society (Beck, 2009), it should, however, not fall into a reactionary form of communitarianism:

The basic mistake of communitarianism is to react to individualization. It is 'reactionary' in its attempt to recuperate the old values of family, neighbourhood, religion and social identity, which are just not pictures of reality anymore.

(Beck in Beck and Beck-Gernsheim, 2001: 208)

A return to communal living as a conservative and potentially regressive reaction to the development of society is seen critically. Its formation and emergence in local communities is indeed a potential answer to the dangers, fears and pressures of capitalistic developments, but, as Castells (1997: 64) reminds us, it comprises, in most cases,

> defensive reactions against the impositions of global disorder and uncontrollable, fast-paced change. They do build havens, but not heavens.

Nostalgia and longing for community must therefore be seen as an ambivalent discursive construction, as a longing for the ideals of being there for each other and the construction of various ways to achieve this. Longing for other times and/or other places does not necessarily mean a critical engagement with neo-liberal developments. Rather, nostalgia itself can also be utilised by neo-liberal politics, processes of commodification and consumerism (Özyürek, 2006; Smith et al., 2011). Similarly, the social critique inherent to nostalgic expressions can situate ideals of solidarity and community exclusively in an imagined pre-capitalist and organic past (Bonnett, 2010). Nostalgia itself can thus not be seen as social critique by definition; rather, elements of critique and protest need to be identified, evaluated and analysed. Nostalgia must be seen as a product of the change in people's living circumstances and at the same time as criticism of current developments. Boym (2001) warns us about the possible negative consequences of an uncritical engagement with an embrace of the past. She describes nostalgia as a sentiment fulfilling particular functions when dealing with and relating to economic, social and cultural processes. While nostalgia can offer potential solutions, it is itself, however, not a productive practice per se:

> Invented tradition does not mean a creation ex nihilo or a pure act of social constructivism; rather, it builds on the sense of loss of community and cohesion and offers a comforting collective script for individual longing. There is a perception that as a result of society's industrialization and secularization in the nineteenth century,

a certain void of social and spiritual meaning had opened up. What was needed was a secular transformation of fatality into continuity, contingency into meaning. Yet this transformation can take different turns. It may increase the emancipatory possibilities and individual choices, offering multiple imagined communities and ways of belonging that are not exclusively based on ethnic or national principles. It can also be politically manipulated through newly recreated practices of national commemoration with the aim of re-establishing social cohesion, a sense of security and an obedient relationship to authority.

(Boym, 2001: 42)

Hence, the reference to and longing for community in the context of care must be evaluated critically. On the one hand, the idealisation of community provides an understanding of what really matters to people in their imagination of care for elderly people. An imagined return to earlier times of family care and community by simply replacing public provision and organisation of care, on the other hand, needs to be avoided. Putting the values associated with community into practice remains an uncertain and doubtful endeavour. In that sense, the nostalgic construction means that community is desired but, at the same time, neither available nor attainable.

The impossibility of community

Nostalgia means that community represents another way of living, which is, due to economic, social and cultural arrangements, neither reachable nor achievable. In the following exchange from a focus group in the United Kingdom, a nostalgic notion of community is expressed, situated within different local and global communities. In the discussion, however, it is made clear that this imagined community is impossible in current times, and that a decline in community is inevitable with societal progress:

Will: I mean, it's this sense of family that the Indian community, and many of the European communities, I've now just been to [European country] [...] but, you know, the hotel we stayed at, there were 3 generations of that family, running that hotel and there was this sense of family, we joined that family, you know, and there's that sense of family, ok, because [...] they've probably lived all their lives, in that little, little town, [...]

Pamela: I actually think that as the generations go on and as people get more and more educational opportunities [...] even within the Indian community, this strong sense of family will...

Will: Yes.

Larry: it will change.

Pamela: it will change, [...] that's right because nothing stays the same.

Morgan: We can't go back to the sense of family that there used to be.

The argument develops that some change has happened and that this change was inevitable. So it can be concluded that nostalgia for a particular idea of community inherently includes a realisation that this form of community is impossible at the moment and, to some extent, unachievable under current conditions. I have already mentioned the emphases on education and economic development, which have led to faster, more individualised life conditions. Bauman (2001: 46) in that context argues that

> nothing endures long enough to be fully taken in, to become familiar and to turn into the cosy, secure and comfortable envelope the community-hungry and home-thirsty selves have sought and hoped for.

Because of the absence of an ideal community, present care arrangements can only ever be a second-best solution. In this sense, one can understand the challenges and difficulties related to the idea of 'care in the community'. Care is closely linked to what is meant to be community but which can never be performed by the 'real' community that is available. Are community and neighbourhood thus images of another life – a life that is desirable and a life in which being there for each other, that is, caring for each other, is possible? Because community is constructed not only as an ideal but also as an impossible ideal, it provides the safety and comfort in opposition to an imposed social reality. People acknowledge that they cannot live up to the ideal (which might be characterised, for example, by living a more 'caring' life) because the ideal is impossible. To some extent, the possibility of personal agency to take over responsibility for the other is challenged by a discursive construction of the impossibility of community. Nostalgia, understood as a yearning for another state of arrangements, diminishes the possibilities for agency in the design of societal arrangements. As mentioned above, the loss of community is often linked to economic and educational advances, arguing for a historical perspective in which community will change within other ethnic groups too as an inevitable consequence of progress and development. It is important however to understand that the loss is itself of the imagination of an idealised situation, of an idealised family and an idealised community. Nostalgia constructs an imagination of a counter-example to the current circumstances and is not (or not only) related to real events and situations. In relation to the family, Beck and Beck-Gernsheim (2001: 129–130), for example, argue that

> the pre-industrial family was mainly a union born of necessity and compulsion. [...] And the strong social cohesion, praised in later times [...] stemmed mainly from an awareness of mutual dependence.

With reference to Benedict Anderson (1991), community is also unimaginable since a complete and detailed sketch of the imagined community would destroy its image. In relation to US discourses, Coontz (1992) demonstrates, for example, how the image of the family has been idealised and a (white, middle-class) myth around this idealised family has been created. However, Coontz (1992: 9) argues that the imagined ideal family of the past is 'an ahistorical amalgam of structures, values, and behaviors that never co-existed in the same time and place'. The mythical construction of communal living and care within and by the community focuses on a past that never existed and is at the same time related to a longing for a concrete and specific experience, imagined as the opposite to insecure and threatening modern times, as Boyd (2001: 8) beautifully describes it:

> Modern nostalgia is a mourning for the impossibility of mythical return, for the loss of an enchanted world with clear borders and values; it could be a secular expression of a spiritual longing, a nostalgia for an absolute, a home that is both physical and spiritual, the edenic unity of time and space before entry into history. The nostalgic is looking for a spiritual addressee. Encountering silence, he looks for memorable signs, desperately misreading them.

In the context of care, nostalgia can thus be described as a longing for a time and space in which ideal care for elderly people was both possible and provided. However, nostalgia, in both its temporal and spatial dimensions, cannot be reduced to a yearning for other times or other places but always includes recognition of the impossibility of this endeavour. Thus, even though this form of care is regarded as ideal and desirable, it is not available and possible in modern times. Nostalgia can therefore be described as an imagination of a utopian place and time, which is however discursively usually situated in the past, as Rubenstein (2001: 4), writing about nostalgia in fiction colourfully, puts it:

> Nostalgia encompasses something more than a yearning for literal places or actual individuals. [...] Even if one is able to return to the literal edifice where s/he grew up, one can never truly return to the original home of childhood, since it exists mostly as a place in the imagination.

The question remains of how far a nostalgic idealisation of community expresses clearly identifiable desires, wishes and solutions for the arrangement of care. The following discussion is a good example for people struggling to find terms, images and concepts for the care arrangements they would favour:

Larry: It [...] works the other way, my late parents lived in [city] and they had fabulous neighbours and we were in [another city] and we used to commute up and down the motorway, but we knew that we could always also ring Andy and Jude if we [...] had a problem, so it is two-way traffic. And of course with this changing society where (...) parents are going to the country, [away] from the kids, and they're moving around.
Will: Well of course, the family unit is totally changing.
Larry: Is changing yeah.
Will: And although I've said, that really we shouldn't be responsible for our parents, the breakdown of the family unit in the UK, [...] that is what is causing a lot of the [...] problems. Because, for instance, I know families in [town] who, for instance, old mining communities, who are still there, 3, 4 generations. [...] The sense of family there is so much stronger than where I live, because they've not moved more than maybe 5 miles away.
Morgan: It's not a breakdown of the family though, is it? Where are **your** children, you know?

Morgan's rhetorical question *where are* your *children* summarises the ambivalence that can be found in the discourse on how care can be arranged and how communal living might be possible. It has been a main aim of this chapter to take the ambivalence in people's ideas, opinions, emotions and experiences seriously. The discussion above continued with a focus on the requirement for mobility in modern society and the labour market and the expectations of and from people who want to succeed in this economic system. Community is thought to contrast with the marketisation and individualisation of modern life; it is not seen, however, as something that can be brought back. Rather, the changing family structures, increased mobility and education, the roles of women in society and other social, cultural and economic developments are often seen as starting points for a new definition of community. The example Larry gives, of the neighbours who looked after his parents, is an illustration of an extension of community beyond naturally grown ties. Beck (1998) describes this experience of longing for community but not wanting to go 'back' as a collective fate, arguing 'no one wants to go backwards. The sacrifice of a bit of hard-won freedom is something that everyone, man or woman, expects only of others' (1998: 34). As has been argued, the nostalgic imagination of community is often related to feelings of personal and social loss. On a broader level, nostalgia expresses some experience (and criticism) of societal decline. A strong focus on public and political discourses embraces the goal to regain the state (of community) we have lost (Sprigings and Allen, 2005). Tate (2007) rightly points out the close resemblance of the discursive construction of community to a feeling of melancholia. She argues that constant 'feelings of loss, and the (im)possibility of recuperation, haunt the talking into being of community' (Tate, 2007: 3). Again, Dench et al.'s (2006) study of developments in the East End of London offers interesting insights into the construction of the idea of a loss of community. They argue that, in particular in economically difficult times, there 'was commitment to local community – involving concern for the needs of *others* – which served you best in the end' (Dench et al., 2006: 47). With economic, capitalist development, the 'need' for community as direct economic support got lost. Community, and how it is constructed, is not a historical phenomenon but a combination of feelings, ideals, wishes and emotions and, at the same time, an expression of being dissatisfied with economic demands and pressures:

> [m]ost individuals still attempt to carve out space for personal commitments, family ties, and even social obligations, but they must do so in *opposition* to both job culture and consumer culture.
>
> (Coontz, 1992: 178–179)

Productive nostalgia

I have argued so far that nostalgia (e.g. for a particular idea of community) includes a realisation that this state of society is impossible and to some extent unachievable under current conditions. The requirements of modern society and the labour market and the expectations of and from people who want to succeed in this economic system challenge the ideals associated with nostalgic imagination. Modern bureaucratic capitalism, the 'most distinct form of Gesellschaft' (Tönnies, 1955: 28), and globalisation have increased this ideological distinction even further. Castells (1997: 60), for example, identifies a clear reaction of people against this development and states that

> people resist the process of individualization and social atomization, and tend to cluster in community organizations that, over time, generate a feeling of belonging.

However, it has been shown that a retreat to community is neither desirable nor achievable either, and hence nostalgic imagination does not necessarily mean that a particular situation is something that can be brought back. In order to identify concrete wishes, desires and possibilities for care arrangements, it is necessary to investigate how far the nostalgia for community can contribute to the productive construction of care. For this endeavour, two aspects or kinds of nostalgic expressions can be distinguished. In the context of nationalist nostalgia, Boym (2001; see also Duyvendak, 2011) proposes a separation between restorative and reflective nostalgia, whereby the former relates to 'anti-modern myth-making of history' through which people imagine a truth that is situated in the past and should be revived. Restorative nostalgia, Boym explains, follows the reconstruction of elements and monuments of the past, focusing on dreams of other times and other places. It is the reflective version of nostalgia that might offer the potential for a progressive intervention into modern social and economic discourses. This notion of nostalgia relates to elements of comparison with unwanted and criticised aspects of modern social living and is thus utopian rather than melancholic (Pickering and Keightley, 2006). The idealisation of community, as well as the idealisation of traditional family structures, acts as a nostalgic sketch of counter-images to recent trends of individualisation, marketisation and commodification. Reflective nostalgia can play a crucial role in the moral and political construction of these images.

Pickering and Keightley (2006) demonstrate that nostalgic manifestations are both product and consequence of a loss of faith in societal progress. People express nostalgic feelings because they resent the social and cultural destruction that accompanies neo-liberal development. In that sense, nostalgia does not only mean a longing for other times and the expression of lost times and societal decline, but must also be understood as an inherent product of change (Boym, 2001). In thinking about the meaning and consequences of a nostalgia for community, Coontz's (1992: 6) argument that myths 'bring [people] together and reinforce social solidarity' might also give some indication of the positive, progressive use of this longing. If times of social solidarity are longed for, community could potentially represent a more progressive way of life in society. For example, community understood as a more flexible and voluntary association of individuals can replace earlier notions of the extended family. This notion of community is detached from a shared history or family ties; it is a notion that emphasises personal responsibility within a neighbourhood as a replacement for traditional social ties. Community care can refer to arrangements in which care responsibilities are shared due to recognition of societal interdependences. General responsibility for each other then becomes the main focus of care arrangements. In the public discourses on care, glimpses of the understanding of missing responsibility can be found as the following short extract exemplifies:

Otto: What's lacking in this society is responsibility.
Silvia: Yeah, I agree [...]
Olive: People don't take responsibility, you know [...]
Fay: I know what you're saying and to an extent I agree, but when it comes to care for the elderly, I don't think it will be that simple.

Fay agrees with the idea that family cannot be the main provider of care anymore. She, however, argues that there will always be and always should be community which can then be understood as the modern expression of being there for each other:

Fay: It's just that society is so different now, but I think that the community, we can only deliver the services we need through the community, and paid for by the state.

I agree with Nisbet when he states that '[c]ommunity forms the ideal sketch of the life that is desirable' and community relations 'come to form the image of the good society' (Nisbet, 1966: 47). And this, I would argue, is an extremely important aspect of the meaning of community in the discourse on care. It sketches an ideal, an ideal society and an ideal way of living, and caring for elderly people represents one part of this better way of living. An uneasy feeling with the current living situation is expressed. This criticism, however, is not directed at oneself; rather, it again expresses an aversion to the economic and social developments in society. Two aspects of nostalgia are particularly important: its relation to and understanding of current social processes and its relation to a critique of social change and developments (Strangleman, 2012). While nostalgia often expresses itself as an 'unloved' sentiment based on a general longing for some attachment, meaning and humanity (Bonnett, 2010), it nevertheless bears important aspects of social criticism. Since the yearning for attachment is based on and emerges from real social processes in which people experience a fragmented and challenging society, Bonnett (2010) calls nostalgic expressions 'necessary myths'. Similarly, Davis (1979) situates nostalgia as a yearning for the continuity of identities in a time when people are experiencing subjective discontinuities. Nostalgia's temporal focus on the past always entails a critique of changes that have happened and is thus a criticism of the situation in situ (Duyvendak, 2011). Pickering and Keightley (2006: 938) show that nostalgia's focus on the past makes it possible for analysts to identify the processes through which the past is constructed, that is,

> according to the imperatives of modernity and late modernity, not as a static, isolated system of representation but as part of a wider temporal orientation whose characteristics are historically grounded and subject to change over time.

Maria neatly describes the need for a new construction of care arrangements, which focuses on responsibility, and not on traditional family and community ties:

> In the days when we lived in small communities care could be shared among all members of the community and/or extended family. [...] But in today's crowded yet fragmented society we need a different kind of care. [...] Society as a whole does have a responsibility – people work all their lives for the economy of the country and deserve recognition and respect when they reach 'old age'. This means a system of support and care regardless of income and need, just as we [support] and care for children.

In nostalgia, Bonnett (2010) always sees the potential for hope and a striving for a different future. Acknowledging the ambivalent notions of nostalgia means to accept the different, often contradictory, desires it entails. Recognition of nostalgia also means a recognition of modernity's alienation and, at the same time, a recognition of the imagination of authenticity related to the past. Bonnett thus concludes that we cannot avoid these stories and tales but should use nostalgic expressions for their potential to 'narrate our journey through modernity' (Bonnett, 2010: 171). In that sense, an imagination of the community of the past can be seen as a potentially progressive notion, which poses an 'alternative to the top-down and top-heavy bureaucracy of the capitalist welfare state' (Brenner and Haaken, 2000: 344). Brenner and Haaken (2000), however, rightly ask how we can distinguish between regressive and progressive elements of utopic nostalgia. While a narrative construction of the past always entails some reference to the present, Alison Blunt's (2003) notion of 'productive nostalgia' moves further by emphasising that this version of nostalgic expression is based on embodied and enacted practice, rather than being exclusively focused on imagination and narratives. In an interesting essay on the topic, Hutcheon (1998) identifies the *irrecoverable* nature of the past as a crucial element of the nostalgic construction. Because of the fact that the past is gone and cannot be reproduced anymore, it holds emotional appeal and associations. Additionally, because the past is inaccessible and only exists in people's imagination and idealisations, it has the potential for people's projections of needs, wishes and desires, from both a conservative and a progressive perspective. According to Hutcheon (1998), nostalgia only picks up imagination related to the past and is thus not linked to real experiences or events but needs rather to be understood as an evaluation of the present. Nostalgia thus makes use of 'retrospective utopia', using the past to express desires and social critique in its engagement with the present conditions. This element of using the past to imagine the future, Boym (2001) describes as creative nostalgia (contrary to prefabricated nostalgia). The focus on the past provides ideas, imaginations, dreams and desires, which help to shape the present and the future. Community understood in that sense can be seen as a counterforce to economic, political and social developments, which people perceive as hostile, individualising and pressurising. Robertson (1995: 30) even argues that because of its counter-movements, 'globalization has involved the reconstruction, in a sense the production, of "home", "community" and "locality" '. However, this redefinition of locality and community also needs to be understood in its nostalgic expression as a search for tools for social critique. We need to be aware, at the same time, that both modernisation and nostalgia should also not be rejected immediately but rather their claims for solidarity should be carefully investigated and evaluated.

> One is nostalgic not for the past the way it was, but for the past the way it could have been. It is this past perfect that one strives to realize in the future.
>
> (Boym, 2001: 351)

An awareness of both past and present is crucial for an understanding of the elements of nostalgia and its potential for progressive interventions. Importantly, the focus on the past should not function as an imagination of a transformation, to create the conditions as they once were (or rather are imagined). A progressive use of the past, what bel hooks (1990) describes in her concept of the politicisation of memory, distinguishes this pure longing for times gone from an active 'remembering that serves to illuminate and transform the present' (hooks, 1990: 147). A focus on the imagined and idealised past and traditions can thus have important elements for a discursive and practical intervention in times of commodification and neo-liberal social arrangements. While nostalgia has often been seen as the antithesis to a radical critique of neo-liberal social arrangements, it actually functions as an important part of the 'radical imagination' by providing ideas, images and desires (Bonnett, 2010).

Conclusion

The demarcation of the meaning of care crucially draws on an imagination of communal living and relating. Community in that sense does not refer to one specific theme or topic of discussion; rather, community represents a combination of interrelated thoughts, emotions, wishes and ideals, as Bauman's description exemplifies:

> To insecure people, perplexed, confused and frightened by the instability and contingency of the world they inhabit, 'community' appears to be a tempting alternative. It is a sweet dream, a vision of heaven: of tranquillity, bodily safety and spiritual peace.
>
> (Bauman, 2004: 61)

Community can be understood as an ideological extension of family, and the neighbourhood in which community is imagined as taking place functions as a constructed extension of the private home. Taking the different but interlinked aspects of the discourse on care together, a picture of the ideal of community can be painted which facilitates security and safety in an ever-faster developing environment. Building on earlier discussions, this chapter has provided another part of the answer to the question of what care means to people and how care is ideally imagined. It has also identified aspects of how the responsibility to *be there for each other* is formed and what community responsibility can mean in the context of care for the elderly.

The meaning of community draws on three intersecting discursive images representing different dynamics of the construction of care. Community as extended family strengthens the values and virtues associated with ideal caring relationships and generalises these values beyond concrete relationships. Community understood as a caring environment builds on the meaning of idealised geographic settings of care. A community then functions as an extension of the own home to include a combination and the interrelation of physical, emotional and ideological connotations and associations. Community, similar to family or home, is the realm that provides safety in a world of market-determined lifestyles. Robertson (1992) speaks therefore of an extension of the ideology of home, one which is not only restricted to the physical space of the house but also includes neighbourhood and community. Like family and home, communities must not be idealised as they can also be (and often are) far from nourishing. Communities, however, are nostalgically alluded to as representing better care, either in the other country/culture where care is *still* family based or in other former times when people *still* cared for each other. Finally, community is thought of in terms of sustainability and growth. In contrast to the abstract and anonymous logic associated with capitalist production, labour and organisation, community conveys an element of organic, natural development.

The discursive construction of the meaning of community draws on a dichotomous contrast between current, undesirable times, in which ideal community has been lost, and an imagination and idealisation of other, past times or other places. The concept of nostalgia shows the processes of this longing and yearning for other, better circumstances in which community was supposed to be possible. The clear dichotomy is important for the discursive construction, as the implication for the rejection of and dissatisfaction with current times only gains meaning through confrontation with an idealised state. Nostalgia allows formulating a particular utopia without the need to specify the practical aspects of these arrangements. However, both empirically (in terms of care arrangements and discursive patterns) and theoretically, there are patterns that extend beyond the strict nostalgic dichotomy. While the nostalgic idealisation of community plays an important role in the conceptualisation of discontent with current arrangements, community in its historic (or rather historically imagined) form is seen as antagonism in current times. Hence, community in its ideal version is impossible and cannot be imagined within the current social, political and economic circumstances. The evaluation of the question posed at the end of the chapter, whether the idealisation of community entails 'productive nostalgia', shows ambivalent results. According to Boym (2001: 355), nostalgia can be called both 'a social disease and a creative emotion, a poison and a cure'. This ambivalence over nostalgia means that it can be found in both conservative and progressive constructions of care, family and community and, at the same time, is rejected by both (Bonnett, 2010).

Pickering and Keightley (2006: 921) thus argue that while the concept of nostalgia should not be dismissed its claims need to be reconfigured. Instead of a return to some idealised past, the potential basis for renewal and emancipatory interventions should be emphasised. Boym (2001) calls for a rejection of a discourse that favours a return to an image of the past and, at the same time, a productive use of nostalgic notions to sketch and imagine changed and improved social, political and economic conditions. How can the nostalgic construction of community be used positively and productively to describe the meaning of care? Which social policies and social interventions can hold onto the productive aspects of a nostalgic construction?

> Between the symmetrical errors of archaistic nostalgia and frenetic overmodernization, room remains for microinventions, for the practice of reasoned differences, to resist with a sweet obstinance the contagion of conformism, to reinforce the network of exchanges and relations.
>
> (de Certeau et al., 1998: 213)

The way new forms of care arrangements are designed and what the new form of community might look like are, inevitably, not classified in clear terms. They involve, however, the attempt to reconcile the freedoms gained from economic and social independence with responsibility for each other. Community thus tries to break out of the traditional dichotomy of security versus freedom and community versus individuality (Bauman, 2001). It is furthermore useful to disentangle two interlinked aspects of the discursive meaning of community described above. Firstly, community is constructed as a unity of people who have something in common, people who share something (e.g. religion, nationality and ethnicity). Care is seen here as a duty between people who share some ties, similarly to the ideal of the (extended) family. Secondly, however, community can be and is also understood as being born and sustained simply and exclusively out of the dedication of its members (Bauman, 1993). Whereas the first notion of community is based on the sameness of its members, and therefore linked to a necessary absence or exclusion of the other (Bauman, 2001), the latter version is based on a recognition of, as Bauman (2001: 150) puts it, 'sharing and mutual care'. This more egalitarian form of community can be described as 'a community of concern and responsibility for the equal right to be human and the equal ability to act on that right' (Bauman, 2001: 150). When community is imagined as the ideal realm and context of care for the elderly, it is important to emphasise the possibility of this more progressive form as the future for long-term care responsibilities. It needs to be borne in mind that, due to the obvious and inevitable absence of the ideal community, present care arrangements can only ever be a second-best solution. In this sense, one can understand the challenges and difficulties related to the idea of 'care in the community'. Care is closely linked to what is meant by community but which can never be performed by the 'real' community that is available. Community is a moral ideal, situated outside the market and based on an idea of altruistic relations between its members (Firth, 2007).

A question raised in this chapter was whether a conception of community can be understood as a counter-discourse to an individualised, economised and marketised world, and I have also pointed towards the positive aspects of a de-traditionalised form of care and responsibility for care (see also Nafstad et al., 2007 in this context). In particular, two dynamics of modern societies are seen as threatening the ideals and values associated with community (Bauman, 2001). Firstly, community is constructed in opposition to economic and social individualism. Political, economic and social developments are an expression of a culture that has alienated people from each other and their community units. An ethic of individualisation as a 'duty to oneself' (Beck and Beck-Gernsheim, 2001: 38) replaces former reliance on family and community. The main problem, Beck and Beck-Gernsheim (2001: xxiv) argue, manifests itself in a society with 'growing inequalities *without* collective ties'. Secondly, the logic of community is contrasted with an exclusively economic logic of society and relating. Community is constructed as a counterforce to developments, which people perceive as hostile, individualising and pressurising. However, ideal care, perceived as taking part in the community and actions being carried out by the community, is not really possible if the very community itself is not available. Likewise, if care within the community is promoted as a way to reduce public services and shift off responsibility to individuals and their families, then the nostalgic construction of community works regressive and disabling. Trying to combine different answers to these challenges I mentioned a tendency in some discussions to comprehend community as a modern answer to the demands of care for elderly people. If family care is not possible (or not wanted), community could step in and take over the responsibility.

The argument over nostalgic expression and the impossibility of community allows at least a partial answer to the questions about the relation between the neo-liberal societal framework and the ideal of care. The socio-economic characteristics of society define how care can be delivered ('real' community care is seen as impossible) and care is situated in the margins of current societal arrangements. In a sense, individual freedom and participation in the market-driven society stand in conflict with what care means to people. But does that mean it is impossible to reconcile independence and care? How are those who need care constructed in the conception of care as an idealised desire of being there for each other? In the next chapter, I will explore the social reality of depending on others in more detail.

5
Who Is Seen to Be Cared for?
The Construction of the Care Receiver

Introduction

'Even though I do it entirely on purpose, it is very much against my will that, repeatedly, every night, I crap my bed.'[1] With this sentence, Dimitri Verhulst (2013) introduces the main character of his book, *De Laatkomer* [the latecomer], who decides to pretend to suffer from dementia in order to allow himself to leave his old life, his family and his friends and move into a nursing home. The comical, yet touching, story demonstrates the limits and boundaries of what is considered a decent and fulfilled life. For someone who is completely healthy, both physically and mentally, to show (and live) the symptoms of dementia, seems to be an unbelievable paradox. For someone deliberately wanting to move into a care home contradicts the imperatives of independence and autonomy, values that are ingrained into our conception of an active and satisfying life. When the book's hero decides to flee from his own life and environment by joining the world of dementia, dependency and care, we can recognise important moral and social associations with illness, old age and dependency. Losing the capacity to remember one's own life story, becoming estranged from the people who were called family, partners or friends and having to rely on others for all everyday activities captures many people's most horrific imaginations about old age. In care discourses, this perception plays a crucial role in the construction of the caring relationship in general and of the care receiver in particular.

In Chapter 2, I discussed the centrality of relationships for the meaning of care. I argued that the idealisation of a family-like relationship encapsulates a particular image of the ideal way of caring and living in old age. Bowlby et al. (2010: 15) demonstrate that our own experiences and understandings of care are directly influenced by the geographically and culturally varying values and beliefs associated with families and relationships. While the idealisation of particular ways of caring is expressed in the depiction of the characteristics of the relationship between the carer and the care receiver, the construction of the meaning of the need for care is strongly related to the person and the image of the person who is to be cared for. The description of care receivers as a distinct group plays an important role in constructing and shaping the 'resources' (Wetherell and Potter, 1992) for the definition of what old age and care mean for people. In the literature, discursive constructions of later life and the linguistics of ageing have become an important part of the analysis of ageing and older people's lives (Coupland, 2009; Rozanova, 2010; Weicht, 2011; 2013). The description of the processes of ageing and different phases in life leads to the life course being sketched as a journey marked by one's varying relationship with care (Bowlby et al., 2010). It has been shown that particular stages in people's life course are described in isolation and discursively linked to specific social identities, such as the one of the care receiver (Hockey and James, 2003). The construction of the older person and/or the care receiver, including a depiction of their characteristics, needs and desires then, in turn, influences the meaning (and politics) of care in general. This chapter will investigate the construction of the care receiver as a crucial part of the caring relationship. Specifically, I will focus on the portrayal of the elderly as the quintessential person in need of care. Old age and the need for care are understood as socially shaped and structured periods in life, which are produced and reproduced by particular ageist values in society (Plath, 2008).

The construction of the meaning of the ideal and social practice of care requires an ontological perspective on the subject of care. This construction of the care receiver is also a discursive prerequisite for any practical policies or interventions that deal with any form of imaginations of client identities (Wilińska and Henning, 2011). Importantly, the person of the care receiver is not fixed but is subject to discursive, political and materialist changes and interventions. This also means that, through the process of care itself, subjectivities are created, shaped and changed. In her study of a hospital ward, Mol (2002) vividly describes the alterations to *being* through practical interventions, arguing that

> *ontologies* are brought into being, sustained, or allowed to wither away in common, day-to-day, sociomaterial practices.

> (Mol, 2002: 6)

Mol thus argues that not only does the meaning of practices and identities change according to different understandings shaped in discourses, but the core being itself changes through certain practices. Personal experience of care can thus alter the individual and social understanding of the meaning of being in need of help (Twigg, 2000a). While ontological understandings are not fixed but subject to change through (discursive and actual caring) practices, particular discursive associations, images and narratives do establish a normative 'age order' in the context of care. In this classification, the older person functions as the quintessential care receiver, and thus as the *other* whose life is distinctively different (and usually considered worse) than the lives of other actors in society. Old age is then usually associated with vulnerability, dependency and a lack of self-determination. Eva Kittay, starting her highly influential book *Love's Labor* with the sentence, 'Dependence requires care' (1999: 1), points to the inherent link between care and the person who is, or who is considered to be, dependent. Passivity and dependency as associations with the elderly person function as moral and psychological categories, which discursively demarcate the passive care receiver from the active citizen (Fraser and Gordon, 1994). Even though personal experiences might suggest otherwise, societal attitudes hold on to a problematic, simplifying and often exclusionary construction of care, being old and needing help.

The significance of the discursive construction of old age and dependency feeds on an idealisation of independence. Independence, constructed as a normative ideal for everyone, penetrates all levels of modern society (e.g. within families and in relation to the welfare state). In public discourses on care, this ideal is combined with recognition of the importance of helping others (Doheny, 2004; Harris, 2002), while dependence is related to a fear of having to rely on others (Latimer, 1999). Being able to live independently from others, and from the welfare state, has become a moral imperative in political debates in modern society and strongly fits the discursive image of market-directed liberal democracies (Chantler, 2006; Latimer, 2000; Misra et al., 2003). Due to this idealisation and essentialisation of independence, a dichotomous construction of the caring relationship construes the care receiver as a vulnerable, passive, dependent person and, hence, as the 'other to the masculine subject of modernity' (Hughes et al., 2005: 265). The idealisation of independence and the fear of becoming dependent will feature strongly in this chapter's analysis of the person of the care receiver and the meaning of old age within the discourses on care.

In order to depict an understanding of the subjectivity of the care receiver, I will first analyse the associations of a particular state of existence with the meaning of a certain period in a person's life. I will show to what extent aspects of someone's identity are essentialised so that the care receiver is reduced to their vulnerabilities and care needs. In the following section, I will investigate the underlying dichotomies and divisions that separate the care receiver from the active citizen. Specifically, the dichotomy of dependency and independence lies at the roots of the public construction of care, underpinning the *othering* of older people and the totalising of people's identities. I will demonstrate the persistence of this fundamental dichotomy by looking at various playing fields on which dependency and the ideal of independence are negotiated. I will then ask whether and how this dichotomy and idealisation can be challenged. Even though Fink (2004: 15) highlights that 'the delivery and receipt of care [...] is a dynamic process in which the lives of both parties are woven together, disrupting any simplistic division between dependence and independence', discursively, a clear separation can be observed. Different degrees of dependency, which will be explored in the following section, are to a large extent avoided and ignored (Weicht, 2011, 2013). I will also focus on possible alternative views of autonomy and heteronomy, which seem to allow an understanding of a dialectical relationship of interdependence (Sayer, 2011). I will explore whether a rejection of dependency as a social construction is useful and desirable. Two influential theoretical perspectives, both of which focus on dependency and independence as moral and structural categories, will be specifically utilised in this section: First, authors of the Disabled People Movement (e.g. Oliver, 1990; Shakespeare, 2000) focus on the societal construction of dependency and argue for the emancipation and support of people with disabilities in order to avoid unnecessary dependency. The position of the writers of feminist ethics of care (e.g. Groenhout, 2004; Noddings, 2003), on the other hand, emphasise the shortcomings of a model of independent beings and argue for recognition of the inevitability of the interdependence of social actors in society. Do these perspectives offer a fruitful challenge to the fear of dependency? An acknowledgement of dependencies might enable an approach that social life is fundamentally defined by interrelated and often, but not always, mutual, dependencies, a conception that fundamentally challenges economic and social ideals.

The meaning of the care receiver

As I have shown in earlier chapters, people's relationships are significant for the meaning of care; hence, the construction of how people relate to each other is a crucial component (and, at the same time, consequence) of the discursive construction of the care receiver. In political and other public discourses, caring relationships tend to be constructed as a clearly distinguishable separation of the care receiver from other actors. The conceptualisation of the elderly person as the care receiver happens primarily through discursive processes, but subjectivities are also shaped during actual caring practices. Mol (2002), for example, states that medical interventions and treatments enact the objects of their concern, that is, the patient or care receiver. In long-term care, the person in need is often identified with an elderly, passive recipient of benevolence and help provided by others. Bowlby and colleagues describe the construction of the elderly person as being characterised by passivity:

> older people tend not to be conceptualised as learning new ways of being, learning new ways of caring and new ways of being cared for – they often are seen as passive recipients of care or, if not passive, as problematic and unruly.
>
> (Bowlby et al., 2010: 85)

Passivity, helplessness and dependency are recurrently associated with the elderly person and thus become defining features of the care receiver's identity. In the following paragraphs, I want to depict the various aspects that contribute to the discursive association of the elderly person with a passive, dependent care receiver.

The elderly

Obviously not all receivers of long-term care can be described as old (needless to say that the ascription of the label 'old' is in itself vague and problematic). Discursively, however, the person who is in need of care is associated with old age. Wilińska (2010) shows that, in discourse, a normative 'age order' is established constructing the older person as the quintessential other, clearly distinguishable from other groups in society. Similarly, Fealy et al. (2011) summarise that public discourses position older people as a distinct demographic group. As elderly people are presented as a particular, largely homogenous group, differences in degrees of passivity, dependency or vulnerability are avoided and ignored (see Weicht, 2011). Ageing itself needs to be understood as a contested and recurrently constituted and reconstructed concept (for various gerontology theories of ageing, see, for example, Powell, 2006). In an inspiring essay, Jean Améry (1968) describes that, commonly, ageing is not understood as a continuous process (which in fact it is), which simply describes time passing in someone's life; rather, ageing is expressed through associations and symbols, which create a dichotomy between the young on one side and an ageing population on the other. Améry (1968) in particular identifies five aspects that mark the shift from young to old and the ascribed identity of the elderly person: firstly, a focus on the past, a life that has been lived and time that has passed replace an interest in the future. Secondly, a discrepancy between the outer, physical self and the inner identity takes place, meaning that getting old causes one to become a stranger to oneself. Thirdly, being judged by others according to societal stereotypes influences one's perception and leads to an internalisation of the gaze of those around one. Fourthly, old age is associated with an ongoing alienation from the world around one, in the sense that the world is thought of as constantly changing while the older person loses their understanding of this world. And finally, dying becomes the defining aspect of one's life. Améry obviously simplified and exaggerated societal associations and ascriptions. In the following sections, however, I will explore to what extent these markers of old age actually feature in the discursive construction of the elderly person, and I will sketch the possible consequences for the meaning of the care receiver's identity.

In many discourses, but particularly in the context of care, elderly people are associated with the past. While their experiences, memories and past actions are narrated and often praised, future events, possibilities and potentials are ignored. Old age is then not seen as a continuous process of time passing but constructed as a marker between those projecting themselves into the future and those identified with the past. An Austrian newspaper article, for example, demonstrates this association of old age with looking back at the past:

> A person who has got old and increasingly frail can look back at his life and it is about being able to spend the last years of life in dignity and with quality.
>
> (*Die Presse*, 25/08/2006)

While the former life experiences are something to look back on (fondly), the future years are not characterised by personal activity but rather marked by a passive state of existence in which one is to be treated with dignity by others. Bobbio (2001) argues that old age marks an end, a final stage, and that it is 'mainly depicted as decadence and degeneration: the downward curve of an individual' (2001: 24). Furthermore, the elderly person's identity is imagined according to their past. This significance of the past manifests itself then in both positive ways, described by Andrews (1999) as an emphasis on history and the memories in someone's life, and negative ways, which Hepworth (2003: 97) depicts as 'regret, a sense of loss, of time passing and of life coming to an end'. In both cases, however, the elderly person's life is one that is to be understood by a focus on the past rather than the future. This characterisation and construction has important consequences for the meaning of the elderly person as the quintessential care receiver. Ingrid's comment in a focus group discussion exemplifies the clear discursive split between the young and the old, whereby the older person's needs and wishes are constructed in a particular way:

> Ingrid: As a young person you don't mind, do you? You're having fun, there are changes, you're travelling, that's great. And the older you get, the more stiff you get, right? The more you have your things, your experiences and you want to have what you've always had. And you want, especially if you're retired, you want to relax as much as possible and not have to worry too much anymore about things that don't concern you anymore.

The elderly person is almost completely stripped off their identity as an acting subject who shapes their own life circumstances. For care, this passive understanding of reducing elderly people to their past means that the care receiver is constructed as a passive recipient of other's care and benevolence. Importantly, the many different forms and levels of intensity of care and support do not translate into a nuanced portrayal of all those in need of support. Rather, one can see how 'the elderly' are often collectively established as a passive entity. In the newspapers, this manifests itself in use of the passive form: for example, *the elderly people (...) are kept as fit as possible by the committed staff* (*Standard*, 29/08/2006). The images that are sketched depict the elderly person as passively vegetating in a state of sadness, as exemplified by a commentary in *Kronen Zeitung* (30/05/2007):

> I think of the many old people who, often alone, without family, lonely and unhappy, remain without support in their houses. Who don't have anyone who organises help for them. It's not that important whether they have 'dementia' in a medical sense. They don't have the energy to keep their apartment tidy, to do the shopping.

Also, trips to the doctor don't happen as nobody is organising those. Let alone a walk or even an excursion. That's all gone.

Here, it can clearly be observed that the focus on the past is inter-linked to a description of older people's needs, wishes and feelings. The elderly person is sketched as a poor, vulnerable person society is sup-posed to care for. This normative construction of older people's identity, which with Binstock (2005) can be called 'compassionate ageism' (see also Friedan, 1993), is also often portrayed as a discursive image of the generation to whom society is indebted. The terms often used refer to *the elderly people who need care, those who have built up the country and made the country what it is now* (*Kronen Zeitung*, 20/07/2007). A subject position which constructs 'the heroes of the past' is based on a dis-tinction between 'their' particular past and 'our' future. In that sense, elderly people commonly receive praise for what they have done, a dis-cursive frame in which their potential for the future is largely absent. Also, in the focus group discussions, an image of those having built up the country is seen:

Ingrid: Here we should then also provide for
Paul: one really should
Ingrid: for the older people, who did really build up this country.

Older people are constructed in such a way that they should be cared for and that it is the task of others to arrange and design particular support and treatment. This is again based on the creation of a clear separation of the time when someone is healthy and independent from the time when someone needs care. This split leads to a strong connotation of 'othering' of those who are very dependent on others, as *they* do not have a life in the way *we* understand it, as exemplified in Paul's comment below:

Paul: She was in [a care home] for seven years, and she hardly recog-nised her daughter then. And she was, so to say, kept alive by law [...] That really wasn't a life in the sense, how we imagine it, or, how we, or what we understand by it.

The construction of who the elderly person is, is used to define what elderly people need, want and desire, supplemented by a focus on the right way to treat and deal with elderly people. As discussed in earlier chapters, this particular treatment is objectified, that is, a particular way of arranging care is constructed as the right and logical way to act which best fulfils older people's needs, wishes and desires:

> Linda: And still it's: I hand you over to another home, I hand you over to care by someone else. I will come to visit on Sundays and that's simply what old people don't want. [...] The feeling of being shunted off, that feeling is definitely there for older people.

Going back to Améry's (1968) discussion of what constitutes the image of the old person, the focus on the past, the imagined passive identity and the definition of needs and desires by others are strongly present in the discourses on care and thus shape the imagined subjectivity of the care receiver. One other, additional, significant discursive feature focuses on illness, vulnerability and physical and mental decline. In this context, older people are being described as difficult, or a strain on other people's well-being, as the following three extracts from different groups exemplify:

> Hilde: But older people, of course, also turn more and more difficult [...].
> Mona: some really become aggressive.
> Hannah: And they can be really nasty and mean.

Both positive and negative characterisations and identity ascriptions share the view that elderly people form a particular, clearly distinguishable group in society, or as Simone de Beauvoir put it, 'either by their virtue or by their degradation they stand outside humanity' (Beauvoir, 1972: 4). The stereotypes and associations related to the elderly person shape the image of the care receiver and vice versa. In the care discourses, the elderly person almost loses their own identity and 'becomes' the care receiver.

Non-person

The identification of the elderly person with the characteristics of the passive, vulnerable care receiver leads to a totalisation of identity. By defining the care receiver in a certain way, he or she becomes a fundamentally different being (Beauvoir, 1972) compared to those leading and shaping the discourse. In Chapter 3, I discussed Marc Augé's concept of non-places to describe the spaces that are stripped of any meaning, history or identity. A similar (discursive) process is happening in the discourses on care and the care receiver. The totalisation of identity can lead to the construction of a non-person who is reduced to their bodily and mental vulnerabilities and deficiencies.

In a comprehensive overview of social theory, Shilling (1993) shows how the body, which represents both physical reality and socially constructed connotations at the same time (Grenier and Hanley, 2007), has become central to the modern person's sense of self-identity. Bodily features thus shape who we are and how we interact with others. Since the care receiver's subjectivity is based on age-related stereotypes and associations, the older body becomes the crucial point of focus. Taking up Shilling's (1993) emphasis of the fact that bodies are not something we *have*, but something we *are* (similar arguments are put forward by Latimer, 2009; Munro and Belova, 2009), it becomes clear that this particular description of the body shapes our understanding of whom the care receiver is. While the young, healthy body (and mind) is usually linked to decent (active) behaviour, the older body is associated with vulnerability and dependency. The ill and declining body represents more than just a body; it is a signifier of a particular identity and self (White, 2009). If the functions of the body are 'failing', the subjectivity of a person is brought into question since, as Latimer (1999) describes, a loss of bodily health is associated with 'loss of activity, productivity, rationality, and self-determination'. Fundamental, unintended changes to the body and its functions, understood as breaks in the narratives of the body (Munro and Belova, 2009), threaten one's personal integrity and identity or, as O'Neill (2009) describes it, the *ur*-trust in our ontology. Shilling furthermore mentions the significance of death for a proper understanding of the meaning of the body. I earlier pointed to Améry's (1968) discussion of death as a significant feature of the imagination of old age. Shilling sees death as the individual's ultimate reduction to one's body:

> the importance of the body in the contemporary age can only be understood fully by taking into account the modern individual's confrontation with death. In a time which has witnessed a decline in the attraction of religious authority and other totalizing narratives in the West, there is a tendency for modern individuals to be left increasingly alone with their bodies in the face of death.
>
> (Shilling, 1993: 18)

Physical and mental decline and the significance of the presence and imagination of death all shape the understanding of the body of the care receiver and thus also the meaning of the care receiver's subjectivity. Since the meaning of care itself is constructed as being closely linked to the body, and giving and receiving care can be seen as the 'embodied manifestation of interdependency' (Bowlby et al., 2010: 38), the description of the elderly body features prominently in care discourses. Bodily impairments are identified and focused on in the imaginations of the situation and the vulnerabilities of care receivers. Rudge (2009), for example, demonstrates how patients with burns in a ward are a sign of our vulnerability, and Mitteness and Barker (1995) show how incontinence is related to constructions of frailty, disintegration and a loss of moral and social competence. Certain bodily features and vulnerabilities are identified and constructed in stark contrast to the moral imperative of a 'normal' healthy body (Oakley, 2007). By being reduced to the body and its impairments, care receivers' (moral) personhood and identity are largely ignored. Aldridge (1994), for example, points out that, in the media, elderly people are often characterised by a 'done-to vulnerability', which portrays them as passive receivers, dependent on others' goodwill. Care (in particular in its later stages) is then often seen as dirty, unpleasant and very intimate contact. Not being able to complete certain tasks oneself is scary and the very manifestation of dependency. The association of certain bodily characteristics with the personhood of the care receiver also often leads to a situation in which people in need of care are infantilised, disempowered and degraded (Bowlby et al., 2010; Hughes et al., 2005; Sennett, 2003). Constructing elderly people as similar to children is a recurrent feature and is often supported by a reassurance that any measures and treatments are done in their best interest. An image is established in which agency is taken away, both in practice and in the conception of the situation. The representation of elderly people as children (or childlike) can have both positive and negative connotations, the latter being exemplified in the extract below:

> Helena: and that's the thing. They take everything as self-evident. Like spoilt children take everything as self-evident, [...] the old people also take everything as self-evident.

In positive accounts, elderly people are continuously described as *lovely* and *adorable* and their actions are infantilised. The rightfulness of a particular treatment is based on the construction of elderly people's identity, as the following exchange exemplifies:

Barbara: And on the other hand, somebody has given me this recommendation and I have also seen that, that you can indeed also scold the ill person a bit, and also the person in need of care [...] and not having to always do everything for them, and having to give in. Because they forget it, that you have told them off, this they forget again anyways, but somewhere, something remains with them, that it isn't entirely fine, what they're doing [...] So you can really once, of course not all the time, but you can really once also have a strict word with them. Because [...] it goes better after that.

Vanessa: and they, I think, like small children, test the boundaries.

Barbara: That's it, yes, that's it!. [...]

Walter: yes, that's it. You can really say, they become small children again. [...]

Adam: and children test the boundaries, how far they can go, but the old people do it as well.

Barbara: or, you also have to tell them then, this is not possible now, and now we do it like this, and so on, but rigorously, because otherwise [...] nothing works. [...] And you're doing it for their good, don't you. You want to mean well, you do mean well.

As shown in the extract above, the infantilisation of elderly people is often linked to an expression of a reassurance that this is done in people's best interest. Agency is therefore systematically taken away, both in practice and in the conception of the situation. Similarly, Larry, in the following discussion on elderly people who are cooking for themselves in care homes, emphasises the necessity to act in this empathic way, in the sense that restrictive actions are performed in people's own interest:

Larry: and I believe, if the person's got dementia or something [...] they can have a kitchen but [...] it's got a master switch that turns the whole kitchen off, if necessary. Let them boil the kettle, anything beyond that [...] they are not allowed to [...] control, in the nicest way possible, in their self-interest.

I argued earlier that the body plays a crucial role in the definition of the care receiver, and the body is the main focus point for the identification of vulnerabilities and necessary interventions. Additionally, however, people's mental capacities and performances (such as communicative skills) are an important component of the construction of the care receiver. One illness, which can be seen as the manifestation of many ascriptions of shortcomings, both bodily and mentally, is dementia. Not only does dementia cause difficulties in people's day-to-day lives, but it also challenges people's notion of personhood and is thus imagined and constructed as a loss of the self (Tanner, 2013). When elderly people are identified with certain characteristics and stereotypes, these ascriptions are constructed as objective truths, and, similar to the discussion of ill bodies, an identification of the inevitable features of the elderly person can be seen. Illness and old age are essentially interrelated with certain behaviours, as the following discussion about an imagined care receiver illustrates:

> Walter: he's always been like that.
> Adam: Yes, yes, he's used to that.
> Walter: he's always been like that. [...] he doesn't know himself differently. He doesn't know at all how mean he is.
> Barbara: No, no, it is also because of the illness.
> Walter: Dementia comes on top of that.
> Barbara: The illness is causing a lot as well. And [...] it changes constantly. I see that as well, especially with people with dementia, one time they're aggressive and really egomaniac. [...] and also blame the others as well [...] and then they fall into misery.

Over recent decades, dementia has featured very strongly in discourses on ageing and long-term care (Johnstone, 2013). Dementia as the quintessential illness related to a loss of self functions as a challenge to what is constructed as 'normality' in society:

> This is not to deny that people with dementia require long-term care if they decline physically and become unable to maintain activities of daily living, but to highlight that people with dementia may be institutionalized because their behaviours challenge the norms expected within their families and wider communities.
>
> (Innes, 2009: 10)

Dementia thus combines many attributes that are feared in the context of old age and requiring care. People with dementia are constructed as losing both physical and mental capacities. Even more challenging, however, and this takes up Améry's (1968) conceptualisation of the fear of old age, dementia is understood as a loss of one's personality, identity, history and social context. Since the meaning of care is fundamentally based within concrete relationships, the loss of those very relationships is particularly important for the understanding of people who suffer from dementia as care receivers (Kaplan and Berkman, 2011; Manthorpe and Iliffe, 2009). In the context of the construction of care and caring relationships, dementia demonstrates that the need for care and support should be understood within a social context and as a challenge for social relationships:

> If the 'self' is a product of social interaction, and the need to preserve a sense of self remains intact, then enhancing dementia care necessitates addressing the attitudes and behavior of those in the person's social environment.
>
> (Tanner, 2013: 161)

Importantly, Tanner (2013) reminds us to confront a biomedically oriented understanding of dementia, which views this condition strictly as a disease involving progressive decline and the loss of self and identity. Apart from the importance of the social environment, people's inability to articulate wishes and desires in the traditional sense requires care to find ways to respect individual lifestyle preferences, choices and identities (Price, 2008). Social interventions need to draw on both a social and a holistic perspective, understanding the clients' life circumstances and the difficulties and challenges for the whole social network (Swenson, 2004). Another increasingly important topic is the question of the end of life. Generally an important field for care, in the context of dementia this topic gains additional significance since people's ideas of the good life are fundamentally challenged:

> Gita: It's like […] at a certain stage in life, whatever illness it is, […] dementia or some other illness, you do then really wish for the patient to be allowed to rest in peace. That's it! […] you don't wish anymore for him to live but, at least that's how I am, that I say, ok, that's enough now.

Dementia (and similar illnesses) challenges our understanding of what it means to have a proper, active life and so becomes a manifestation of vulnerability and passivity. Living with dementia at home and, ultimately, dying at home requires support for people's partners and families in the process (Hughes, 2013). The main challenge for care in relation to both dementia and the end of life is to 'uphold their personhood whatever the degree of dependence, and facilitating them and their carers in making and operationalizing choices throughout this final phase of life' (Holloway, 2009: 721). Not only in relation to dementia, dependency (and the challenges associated with it) plays a crucial role in the construction of the care receiver. In order to try to disentangle crucial dichotomies, in the next section dependency and passivity will be discussed as defining features of the care receiver's subjectivity.

Dichotomies: Dependency vs. independence

Care is situated within and shaped by close relationships, and caring relationships are, likewise, based on dependency relations. This means that a dichotomy is created between the active carer on the one hand and the passive, vulnerable care receiver on the other, or, in other words, care is constructed as polar opposites of dependency and independence (Chantler, 2006). Conceptually, the care receiver is dependent and relying on the independent actor, the carer. Care in this sense can be understood as the often dirty, unpleasant and intimate reaction to dependency. The dichotomy between dependency and independence reproduces stereotypical assumptions and constructions of societal actors. In the context of disability, Shakespeare, for example, identifies the construction of a 'polarity between dependent, vulnerable, innocent, asexual children and competent, powerful, sexual, adult citizens' (2000: 15). The dependent person functions as the other who is, by way of comparison (Strathern, 1997), used to define what a reasonable life is; the other is hence used to define one's own independence. The infantilisation and othering of people are crucial markers of the construction of the care receiver and are, conceptually, based on this dichotomy (Oakley, 2007). In this section, I explore the ways in which dependency is constructed as the *other* way of existence, one which is prefigured as undesirable, feared and rejected.

Dependent vs. independent

Public discourses demonstrate that the way that care is talked about reflects a dichotomy of independent and dependent actor(s), with one person being active in the process and another being passive and dependent on the former. Lloyd (2004: 247) defines care as taking 'into account the needs and rights of those too young or too old to be the active "independent" adult citizen'. Care is thus imagined as the necessary reaction to the individual shortcomings of those who are dependent. A focus on dependence inevitably implies that there is a group of people who lead independent lives and another group who do not or cannot, with the latter group requiring help from the former (Johnson, 1990). Independence becomes the ideal against which care and support needs are measured.

Independence is a persuasive concept and can be seen as a dominant paradigm of current political and social policy arrangements (Fraser and Gordon, 1994; Mittelstadt, 2001) and social, political and economic life in general. Independence as one of the key characteristics of a good social life (Harrefors et al., 2009) is very strongly integrated into people's conceptions of their own existence and societal structures, or as Oldman (2003: 45) puts it: 'It is almost impossible to contest the concept of independent living, as it is hard to challenge motherhood and apple pie.' Originally mainly an economic classification (in relation to work enabling or preventing independence), Fraser and Gordon (1994) describe the rise of dependency as a moral/psychological and thus normative category, meaning that those dependent on others are constructed as being morally inferior to the idealised independent person. Independence and dependence as ideals and principles have a long history in philosophy (Held, 1990), economics and policy-making, but also in everyday public and social discourses. Fraser and Gordon (1994) show that the concept of dependence has substantially changed over time, fitting the ideological and political demands of respective systems. Especially since the Enlightenment, the values of independence and rationality have led to an idealisation of the autonomous male, white, middle-aged, able-bodied subject as the quintessential social actor (McNay, 1994; Shakespeare, 2000; Watson et al., 2004). Work and employment, which used to be a sign of depending on somebody else, now represent one of the few possibilities for escape from dependence (Fraser and Gordon, 1994). General societal structures and life circumstances define the ideal citizen as an independent, self-reliant actor while any divergence is seen to be inferior to the ideal. There have always been groups of people (women, slaves, the young), however, at whose expense this independence has been created. Additionally, there have always been groups of people who could not embody the ideal of independence due to their position in society or their conditions of life. In the context of long-term care, independence has become the crucial aim of policies and interventions. Glendinning (2008), for example, shows the relevance of ideas of individual choice and self-reliance in more and more fields of care and other social policies. Enhancing the elderly person's independence (Harlton et al., 1998) and enabling 'independent life styles' (López and Domènech, 2009) are seen as political and societal targets to strive for. Dalley (1996) highlights the importance of economic independence (which is often linked to the possession of a property) for any concept of independence. Economic independence gives other principles meaning. Morgan, in the following statement, also points to the fact that people's independence is related to having their own house, and thus moving into another form of accommodation, such as a care home, is synonymous with giving up one's independence:

> Morgan: I also think that the point you've made about the grey pound is right and I think economically, despite the credit crunch and everything we hear, the older generation are better off than they were previously and obviously more people own their homes, and are more independent in that way. [...] So there's an awful lot to give up for them.

While economics is undoubtedly shaping all possibilities for societal relations, independence as a value and an ideal has moved far beyond the merely economic. The societal imperative of individual fulfilment and success has led to a situation in which being and remaining independent is a marker of one's societal role and status. Being independent and, thus, being able to be a person who can, potentially, take care of others is the only desirable state. In care discourses, this manifests itself in the fact that people often express that they want to make the decision to care and to be there for others (representing the heroic, sacrificial, selfless virtue of giving), but that, on the other hand, they do not want to be in need of care (the horror of taking). In other words, material independence is accompanied by moral constructions of self-reliance and self-sufficiency to form a moral ideal of personhood. Dean and Rogers (2004) found some (ambiguous) awareness of interdependence in their interviewees' accounts and '[p]aradoxically, [...] even those who denied their own interdependency demonstrated an acute awareness of the necessity that others should depend on them' (Dean and Rogers, 2004: 74). Related to the moral ideal of not wanting to be dependent on others is the fear of becoming a burden. The following focus group exchange exemplifies this fear:

> Alfred: I have in that sense thought about it, [...] that I say, right, I don't want to get that old. If it doesn't go anymore, then away, away, away. And there, I think, I'm agreeing with my mother. She also ran until the last moment. And then, when it got critical, she gave up. Then she died within 2 days. [...] So, of care she was

horrified, that she will be in need of care, that she would be dependent on other people. And I do understand that very well. I don't
want that either. Then I rather want to go before that. [...] So, care,
no, then I'd really like to go.
Britta: I mean there certainly is something worse than dying and
that's suffering.
Alfred: It is worse if you are in need of care. Yes. It's worse, for
example, if physically you can't, but mentally you're fully aware.

The theme of burden reflects a topic that is very characteristic of the
discourse on care (Shakespeare, 2000). In much of the care-related literature, a focus on the burden on carers has been able to point to the
hidden work that is being done in an often unacknowledged and under-
valued way (Shakespeare, 2000). Social policy-making, which focuses
strongly on the needs and the situation of carers, can similarly lead
to an emphasis on the idea of burden, as Hughes et al. (2005: 261)
argue: 'Social policy constructs male and female recipients of care as a
burden and a drain on scarce resources.' This emphasis on burden in academic, political and public discourses obviously reproduces the negative
connotations of dependence and the negativity appears as an objective problem, since somebody needs to carry that burden. The paradox
of people valuing caring, and even expressing their willingness to care
for their relatives, friends or neighbours while not wanting to be cared
for by anyone close to them, is based on this fear of being a burden.
Being dependent on others is thought of and constructed as a fundamentally passive, negative stage of existence, as Sennett (2003: 63–64)
argues: 'Care of oneself can mean additionally not becoming a burden
upon others, so that the needy adult incurs shame, the self-sufficient
person earns respect.' Care is seen as a pitiful stage in life, a situation
people would like to avoid at all costs. The following extracts give an
impression of people's wishes for their own old age:

Pamela: I would never ask my children to look after me.
Will: No, and I've made that clear.
Pamela: I know that very clearly and that's very clear because I just
don't think it's fair, I only like to be near them, so that I could see
them occasionally...

Dependence is a sign of not being healthy, of being passive, of not being
self-reliant and not functioning as a 'proper' person in society. Modern
society is an arrangement of/for able-bodied individuals (Oliver, 1990)

in which those who are a burden are *the other*. Johnson (1990) therefore calls the ascription 'dependent', a label that marks someone as deviant, as 'someone who no longer enjoys a place in the mainstream of society and whose behaviour is "abnormal"' (Johnson, 1990: 216). In a study analysing people's perceptions of dependence and independence, Dean and Rogers (2004; see also Ellis and Rogers, 2004) report that dependence is generally seen as a negative state that people try to avoid and which is seen as only acceptable for some members of society (e.g. people with certain disabilities), but not for others. Personal responsibility for making specific choices is seen as the main factor to avoid dependence. Imagining oneself through the narrative of care in a situation of helplessness and dependence is thus a very negative idea for many people. Being in need of extensive care is described as something different from life *as we know it*.

> Nathan: My prayer is that I keep healthy, till the day I die [...]. And I never need looking after.

People express their unwillingness to live a life of *being a burden* and effectively of being dependent on others. This leads to a wish to die healthily without needing any care, but it also includes a discourse in which euthanasia is repeatedly thought and talked about. Ingrid, for example, declares that she would consider ending her life in response to a discussion on the possibility of being in a care home:

> Ingrid: If I'm able to, to do it, to understand my situation, and I can't see a way out, then I would make use of my right, to determine the point of my death.

This dichotomy between dependence and independence thus distinguishes between full active members of society and those for whom care and support need to be arranged and provided. The identification of the care receiver as the dependent, vulnerable person emphasises their passivity and helplessness. Obviously, not all elderly people are passive and dependent. Hence, a further dichotomisation takes place in discourse, differentiating between active and passive elderly.

Active vs. passive elderly

In the previous sections, it sometimes seems that I equate old age and dependence within caring relationships. This is obviously not a reflection of reality and many older people do not require care or are even caregivers themselves. However, this apparent observation entails another significant dichotomy that shapes our understanding of the care receiver in discourses. The separation of those older people who are *still* seen to be active from those who exemplify the passive care receiver reproduces the binary construction of dependence and independence. In the sketching of older people in discourse, this seems to be a recurrent pattern. Wilińska (2010), in her analysis of the Polish context, identifies three positive roles for elderly people: the grandparent, the rich pensioner and the person who stays active in the labour market. All other identity positions are characterised by and associated with negative connotations of dependence and passivity. Similarly, popular current identity constructions related to successful and/or healthy ageing can reproduce and reinforce the split between those who are active and those who are passive elderly care receivers (Gilleard and Higgs, 2011; Liang and Luo, 2012; Rozanova, 2010). In the discourses on care, questions of activity and passivity are important descriptive markers of the identity of the dependent care receiver. In the following extract, Helma talks about the act of caring itself, describing how an old person can help in the process of caring. This is a good example of the idea of 'othering' (Oakley, 2007) old people, who lose the identity as agents in their own right:

> Helma: Because I really believe [...] and I also know old people who are like that, adorable, quiet, grateful, not that they are now dismissive [...] but dignified grateful, yes, that you like to have them with you and that you like to be there for them [...] Now there are those and then the whole caring would be much easier because they simply would also be there.

Elderly people in this context are presented and talked about very fondly, in relation to an emphasis on their adorability as ideal patients or receivers of support and help. In contrast, the active older person is praised for their fitness, be it mental or physical. Newspapers run stories that emphasise (some) elderly people's fitness and activity:

> Nowadays, people beyond the age of 60 don't feel like only being one of the old guard. They are active, both physically and mentally.
>
> (*Kurier*, 29/06/2006)

Such contributions describe and praise elderly people's involvement in and contribution to work, sports, learning, studying and volunteering.

Similarly, in the focus groups, people sketched a picture of some elderly people who are seen as examples of active involvement in society. The active older person functions as the quintessential opposite to the older person associated with care. Eloise, for example, talking about respect for elderly people in general emphasises how great are old people who are still active and intelligent:

> Eloise: and age itself, now, that's not really a merit. But, I mean it's great, I know many old people, who I also like, where I think that's great, how they are still interested in things, and how smartly they can talk, but age itself, that's not it.

Similarly, in the following extract, people who deal with elderly people in the context of voluntary work emphasise how some are very active, funny and witty. Highlighting that someone who is in their late nineties is active and still going on holidays, the discussants present an ideal of an elderly individual, which is the opposite of a dependent, passive person.

> Peter: This is very important because there is some prejudice against elderly people and invalided people but, people tend to walk past and they are probably fully aware insight their brain. Fully, with a 20-year-old brain [. . .], 30-, 40-year old, instead of a 70-year-old person. [. . .]
> Patricia: Our favourite lady at the moment is 98.
> Nathan: She said she wants to go to Australia.
> Patricia: next year.
> Nathan: For a holiday. At 99.
> John: It's amazing.
> Patricia: And she came in the other week and she said 'Patricia, I'm a bit worried', I say 'Why [. . .], what's the matter?' She says: 'I'm beginning to feel my age' [laughing]. And I: 'I'd worry about it when you start acting your age' [laughing], 'cause she doesn't act her age [. . .]

It is important to understand that it is this emphasis on older people's abilities, actions and cognitive skills that reproduces the dependency–independence dichotomy. Due to the emphasis in elderly people still being fit and healthy, the negative associations of dependency are perpetuated. Betty Friedan (1993) favours the portrayal of active older people and warns against the association of old age with dependency:

Thus, to base our main image of age on nursing homes is to reify – and even make a self-fulfilling prophecy of – that terrifying image of incontinent, senile age.

(Friedan, 1993: 16)

Problematically, however, while the intention is certainly worth supporting, this warning also reproduces the very fixed negative meaning of dependence and care needs. In the literature, the separation of active and passive elderly people has been conceptualised as a distinction between the third age and the fourth age (Gilleard and Higgs, 2011; Whitaker, 2010), whereby the former is characterised by activity and societal involvement and the latter by shortcomings and dependence (Baltes and Smith, 2003).

Bringing together the idea of active elderly people and the fear of becoming a burden is a discursive feature that emphasises older people's contributions. In this context, people refer to the worth of elderly people in terms of their involvement in society and their value to others. While not contributing anything leads to *feeling useless* and is seen as *a very heavy burden*, elderly people's contribution to others, in relationships or families, to the local community and to society as a whole is praised (see Wilińska and Henning, 2011):

Adam: And of course, in times when the family, from the great-grandfather to the great-grandchild lived together, usually in a farmer's house, there was also the advantage that the old people then at the same time cared for the children. Today they have to be put in a kindergarten, and you have to pay for it, or the taxpayer has to pay for it [...] because you go to work yourself. [...] And then it was like that, that the less frail older people, they [...] read to, or sewed something, or washed [...] so that everyone had his tasks, and so contributed to the family's well-being.

While differentiations between those who are active and those who are passive can reproduce the underlying dichotomy of dependence and independence, several empirical accounts discussing concrete relations within families and within society challenge and question the consistency of the dichotomy. In exploring those instances further in the following section, I will search for theoretical positions that allow and foster a questioning of the construction of the binary.

that passivity is thus not an inevitable feature of old age and a need for care. I agree with Phillips (2007: 135) when she argues that '[k]eeping older people "independent" has been translated into a message of keeping them fit and active whereas it should mean giving people more choice and control over their lives'. However, I would add that, given the way in which the concepts of choice and control are used in public and political discourses and policy-making, they are also idealised aspects of an imagination of independence (see Glendinning, 2008). Taking up some themes discussed in relation to community, rejecting dependency can therefore also be linked to a critique of societal demands, facilitated by the logic of the market. The need to be able to participate fully in the labour market and/or as a consumer is challenged and questioned when the logic of dependence is rejected. Dependence on others, and thus the counter-logic to success and economic contribution, is doubted as essentialising and restricting:

> Walter: the old ones do not count anymore, the youth is the important thing. The youth is the future.
> Adam: yes.
> Barbara: Yes, the youth.
> Walter: the youth is the one that brings the money.
> Barbara: Yes, everything's calculated like that.
> Walter: The youth is the one that works and the old guy, whatever, we also have to provide for him.

Again, own choices and decisions are presented in opposition to an image of old age, in which elderly people are dependent, passive and vulnerable to abuse, neglect and infantilisation. The empirical examples of discourses reproduce the idea that old age is closely related to being completely dependent on the goodwill of some abstract other, which through interventions of the other activity is still possible and desirable. On a personal level, planning, choosing and making conscious decisions are constructed as key tasks for someone to be able to escape from the trap of dependence, as illustrated in the following newspaper commentary by a woman imagining her own old age with some friends around her:

> It wasn't just the sociable nature of the enterprise that appealed [of moving into a home together]. It was the thought that we were going to be one bunch of oldies in charge of our destiny. We will choose everything from menu to morphine and thumb our nose at the cruel

convention that the elderly are to be treated like children. We will show the curious visitors that you can be in your sunset years and still be interested in the news or in seeing a good play; we'll exchange views on the Booker shortlist and Gordon Brown's long-term future. No one will take our savings, jeer at our incontinence or nod-nod, wink-wink at our sexual frolicking.

(*The Observer*, 09/09/2007)

While actively rejecting the label and the logic of dependence seems to be a useful discursive intervention, this strategy often means that different levels and states of dependence are ignored. When Friedan criticises the societal portrayal of old people as reductive and one-sided, where 'only the "problems" – the *truly* helpless, dependent, sick, isolated, senile ones – are actually seen as "old" ' (Friedan, 1993: 28), she inadvertently reproduces the normatively charged dichotomy of dependency and independence. One influential theoretical position challenging the equating of elderly people with passivity and dependency emphasises practices of support and help, rather than care, in order to make it possible for everyone to function as an independent actor (Ryburn et al., 2009; Smith, 2005). While the discussion so far has followed the traditional conception of care as a response on behalf of those who have the capacity to respond to the needs of others (Groenhout, 1998: 177), the Disabled People Movement (DPM) advertises the replacing of a notion of care by the terms 'help' and 'assistance' (Shakespeare, 2000), seeing care not only as an answer to but also as a cause of dependence (Hughes et al., 2005). Following the classic liberal rights model of promoting independence for everyone (Ellis and Rogers, 2004), the terms 'help' and 'assistance' carry a meaning of an emancipatory concept of independence and self-determination (Hughes, et al., 2005), whereas care has the tendency to objectify and construct dependent people (Shakespeare, 2000). Drawing on a social model of disability, the main aim is to reject notions of pity and victimisation (Smith, 2005) and instead focus on emancipating and empowering people. John, in the following quote, discusses a photo of a young man fitting new light bulbs and an elderly man standing beside him:

John: Yeah, and the greatest thing that a carer can do in that situation is actually [...] not lord it over them [...]. So actually they're not invading his privacy. He's still taking part, and inviting him, even to hold something, [...] 'can you hold that for me', [...] just not exclude them from the activity but make them feel as though they are part of it, and in fact, that they're the boss.

One problem arising with the concept of empowerment is the agency of the assistant involved in the process. Oliver and Barnes (1998) argue therefore that being empowered by someone else is an inherent contradiction and ultimately represents disempowerment (see also Smith, 2005). A support perspective also assumes that people are in a position to make certain choices during their lifetime to avoid dependence at any stage in life. The emphasis on independence and self-determination has led to a focus on the empowerment of people and participation in public life. As much as these aspects are important in many cases, they also bear the danger of focusing on traditional connotations of what independence means. Harris (2002: 277–278) rightly argues:

> Whilst these strategies are important as a way of countering a view of older people as a passive burden, they inadvertently reinforce a concept of citizenship which defines people's status according to their contribution to the economy, as well as reinforcing a sharp distinction between the young-old/old-old and the grey pound/grey drain.

Another problem arising might be that the focus on help instead of care can reproduce the dichotomy between independent individuals and dependent recipients of assistance even further. In other words, does the focus on help potentially construct those who really need *care* as different to those who simply need help and assistance? It could be argued that a mutual relationship between people is sometimes simply not possible or even desirable (see also Fine and Glendinning, 2005).

Everyone is (inter)dependent

Apart from rejecting dependence altogether, one strategy to challenge the dichotomy is to recognise the existence of mutual dependencies, starting from the question of whether anyone can actually live 'independently', that is totally autonomously from everyone else. In fact, we are all always dependent on others; dependence is inherent in any social being, and it is, as Kittay (2002) argues, an integral part of human existence and human nature. Shakespeare (2000) draws attention to the fact that there are no two distinct natural categories of dependence and independence; rather, the reasons for dependence are inherent in human life: 'Everyone is impaired, and all people have areas of vulnerability' (Shakespeare, 2000: 9). Fine and Glendinning (2005) point out that there are different aspects of inevitable dependence, such as economic, physical, emotional and political dependence, and Groenhout

(2004: 10) portrays individuals under the basic premise of an acknowledgement of necessary dependence as 'physical beings who live lives that are inescapably structured in terms of dependence on other humans and on the environment in which they live'. Mutuality and interdependence are therefore seen as constitutive of a moral being, and being able to be a helpful, contributing citizen is seen as the counterexample to the elderly, dependent person, as the following exchange exemplifies:

> Lisa: I think it's very difficult to preserve a person's dignity and sense of worth when they become a bit helpless. [...] And I think [...] they need to be aware that they are treated with respect for them and for their dignity and their pride in as gentle a way as possible. And in an understanding way. And I also think it's important to help them keep their sense of purpose in life. When so much, gradually, goes from them. [...] That they need to feel wanted
> Carol: And a helpful citizen

Since society, especially in its meritocratic conception, is constructed in a way that favours the individual contribution to the whole, it is the meaning of the idea of not being a full, contributing member of society that makes feeling dependent such a negative experience. Being passive, receiving and dependent is an unwanted, inferior status to have to face. In the conversations, people continuously talk about the worth of elderly people in terms of their contribution to society and their value to others. In the following extract, the importance for one's feeling of self-worth as a consequence of *fulfilling tasks* is emphasised:

> Barbara: here [in town] there was this elderly home, and even then, I don't know, 200 years ago or longer [...] people had to do simple tasks [...]
> Vanessa: yes, that just keeps fit
> Barbara: Yeah, and this was actually very smart, very intelligent from this founder who had ordered that they are asked to do simple tasks [...], that you are not feeling useless. Because that's a very heavy burden
> Barbara: a psychological burden, I think
> Adam: That was of course also the advantage when the family lived in one union, from the great-grandfather to the great-grandson, usually in a farmer's house, that then the old people could at the same time look after the children.

Not only the focus on home but also the general construction of a preference for independence might cause substantial difficulties for people. Dalley (1996) argues that it is the construction of the capitalist society that constructs 'those who cannot work (for wages) through physical or mental impairment, or those who have passed beyond the age limit imposed by society on the end of working life' (1996: 98) as dependents. Whereas I agree with her argument that the social construction of dependency needs to be challenged, I think it is equally important to recognise that for some people the category of independence is not useful in describing their current and future life situations. Eloise, in the following example, also seems to reproduce the aforementioned dichotomy. She, however, positively refers to a character trait that makes the acceptance of dependence possible. In talking about her mother's approval of the situation in the care home, she refers to her upbringing and the fact that economic success and productivity were not the main determinants of life:

> Eloise: the period before that [...] even though she was in a care home, my mother has always been [...] a satisfied type, yes. So, because of war and these things [...] and then you just didn't have that much, and generally she also didn't have such huge demands, and she was also always very grateful, yes. And this has actually then, in the care home, also worked really well.

Conclusion

Demands for being independent, acting autonomously and not having to rely on others feature strongly in modern culture and society and are thus not a minority concern for disability studies or gerontology, but an issue for mainstream social sciences (Barnes et al., 1999). In earlier chapters, care relationships have been identified as the crucial focus for the construction and meaning of care. Within the care relationship itself the care receiver is imagined as the clearly defined dependent, a passive recipient of help, support and care. Even though people show some awareness of the fact that the care status can be ambiguous (Bowlby et al., 2010), the dichotomy of active carer and passive care receiver remains a persuasive feature. Since caring discourses construct 'the elderly' in their role as care receivers in a particular way, this construction in turn forms the basis for the demand for particular types of care arrangements (Wilińska and Cedersund, 2010). In that sense, this chapter's analysis also needs to be read as both a reason for and a consequence of earlier discursive aspects, such as the meaning of home, community or relationships.

The chapter has shown the persistence of the values of independence and self-reliance and the consequences for those involved in caring relationships. Care is constructed as a dichotomy of dependent/independent actors where the image of the elderly person is characterised by dependence and vulnerability. Ideologically, there is one person active in the process, the independent actor, and there is another person who is dependent on the former, that is, the dependent, passive receiver of care. Illness, frailty and loss of capacity are all seen as markers of a general state of dependence and loss of individual autonomy. I have argued that due to its challenge to both physical and mental capacities, dementia functions discursively as the quintessential illness of dependence. In addition to the illness' challenges to both body and mind, dementia is feared because of the threat of losing awareness of one's history and, even more significantly, one's relationships. While the symptoms of dementia and the consequences for both people experiencing them and their carers should certainly not be belittled or ignored (Innes, 2009), it needs to be stated that such fear also relates to the discursive construction of illness, vulnerability and dependence itself. Undoubtedly, vulnerability of the body and mind, and suffering, do exist and create real difficulties for people and should not be romanticised (Groenhout, 2003; Sevenhuijsen, 1998). What is needed, however, is a recognition of the discursive processes that totalise illnesses, vulnerabilities and dependence in a way that fundamentally identifies a person with their role as care receiver (see also Garland-Thompson, 2005). Resisting the totalisation of dependences might mean that

> to treat someone in a dignified way is not to ignore their vulnerability and dependence on others, but rather to treat them in a way which discreetly acknowledges that vulnerability without taking advantage of it, and to trust them not to use their autonomy in a way which would take advantage of our own vulnerability.
>
> (Sayer, 2011: 203)

Linking this discussion to the construction of home, described in Chapter 3, it can be shown that all the negative attributes about dependence are combined and thus manifest themselves in the idea of an institution and institutional living. Institutionalisation therefore becomes a stigma (Phillips, 2007) for those living in it, while at home, on the other hand, this stigma is not apparent as, regardless of the actual situation, independence is potentially possible there. The loss of independence is combined with a deficit of subjectivity or individuality, and the care home is the ideological manifestation of this loss. The unfamiliar setting results in even more dependences and thus works in a disempowering way (see Phillips, 2007). Losing one's subjectivity is sometimes described as a quasi-death, or social death (Froggatt, 2001), and this association is, furthermore, strengthened by a clear relation between moving into an institutional setting, a shortfall of identity and nearing the end of life.

This book's attempt to situate the ontological not in different types of people (as in trying to explain and describe how human beings *are*) but in relations with each other also allows an inclusive and open approach to the emancipation of disabled and older people. Kittay (1999) and Feder and Kittay (2002) make it very clear that care and dependence do not inevitably lead to subordination. Rather, they see the challenge as lying in how society can 'deal justly with the demands of dependencies that constitute inevitable facts of human existence, so that we avoid domination and subordination with respect to care and dependency' (Feder and Kittay, 2002: 3). In this chapter, I have also discussed an important paradox and ambivalence emerging from the discourse on care for elderly people: people want to care and be there for each other, and value (in emotive terms) caring and being there for each other. People do not, however, want to be cared for and do not want to be dependent on others. This links to an understanding of care as a heroic action, emphasising the virtue of giving, whereas being cared for is passive and linked to the horror of taking as a passive recipient. Sennett (2003) identifies giving as a better virtue in its earning of others' respect than being an independent, self-sufficient actor. As dependence is seen primarily as a problem and as something negative, the definition of dependences becomes very important in the public discourse. Complete independence and autonomy need to be seen as an illusion that is based on a particular construction of certain social relations. Kittay (1999: 183) therefore emphasises the importance of other people for the very possibility of independent citizens and workers:

> The purchase price of independence is a wife, a mother, a nursemaid, a nanny – a dependency worker. Whether the care of dependents is turned over to a woman with whom one shares an intimate life or to a stranger, unless someone attends to the dependencies that touch our lives, and inevitably touch the lives of all, we cannot act the part

of a free and equal subject featured in the conception of society as an association of equals.

The fear of dependence thus moves beyond the mere individual and becomes a social challenge and topic of struggle (Johnson, 1990). Reproducing the dichotomous image, care is in a societal context constructed as a practice that is carried out by *us* (as active citizens) and *for* others (those requiring help). Morally laden discursive constructs shape the actions of everyone in society, not least social policy (Rudman, 2006), and social policy plays a part in the definition of social issues (Biggs, 2001). Plath (2008: 1364) criticises the fact that the dominant understanding of independence 'places the emphasis on older people doing things alone and making decisions alone' and argues for an understanding that also includes alternative aspects such as access to community. She also argues that the traditional focus on 'doing things alone', 'making one's own decisions', 'physical and mental capacity', 'having resources' and 'social standing and self-esteem' (Plath, 2008) can all have negative consequences. Oldman (2003) also bases the difficulties of independence on a societal level and criticises the state's complicity in the process of construction. Bauman (2001), for example, has shown that those who are seen as very successful and productive in modern society are also characterised by independence and self-reliance. This link between (economic) success and independence also leads to a moral rejection of the dependence created by discourses shaped by the successful and powerful (Bauman, 2001: 50). It seems that the dichotomy between dependence and independence also reproduces the notions of success in the market versus a life based on relations (Zelizer, 2005).

In this chapter, I have addressed the discursive construction of independence and dependence and I have presented the need for an ethical position which allows dependence (see also Weicht, 2011). At the same time, I want to argue that empowerment, justice and equality are extremely important aspects of a progressive ethics of care. An ethics of care, which, in Fraser's (2003a, 2003b, 2005) sense, gives rise to recognition, redistribution and political participation, also needs to understand the meaning of categories such as freedom and autonomy in the context of specific societal arrangements. In the next chapter, I will explore further the meaning of these categories in relation to care within the social reality in which people live. A rethinking of the processes that shape the societal meaning and ideological construction of independence and autonomy is crucial for Groenhout's (2004) image exchange and the commodification of care. The intrusion of markets and their logic in the care sector has important consequences for those being cared for, and in particular for those caring informally. In public discourses over recent decades, informal care has become a prominent issue. Politically and socially, carers have been praised, and in that way the foundations of this form of unpaid care have been reinforced. Referring to the discrepancies between moral appraisal and economic, social and cultural support, Hochschild (2003a: 2) argues that '[i]deologically, "care" went to heaven. Practically, it's gone to hell.' Ungerson (1999) points out that the academic discussion around care has presented it as being in opposition to work and argues for a breaking down of the boundaries between care and work. She thinks that 'marketisation, privatisation and consumerism have been locked into a symbiotic relationship' (Ungerson, 1999: 585) and somehow created a dichotomy between paid and unpaid care(work). Voluntary commitment and informal care are praised and valued highly, not only for the importance of the work done for individuals and society but also for their characteristics as morally significant work. I have pointed out already (see, for example, Chapter 2) that care is not primarily understood as the fulfilment of a set of divided tasks; rather, it is a complex relationship between the person in need of care, the carer and the environment (such as the community or society). The carer is referred to not as someone providing certain services, but rather as *being the carer*. The focus thus lies on the identity of a caring person rather than on the fulfilment of tasks, the delivery of services or the provision of a certain form of labour.

In order to understand the discursive discrepancy between the idealised physical and emotional presence of the informal carer on the one hand and the increasing influence of marketisation in the sector on the other, this chapter will zoom in on dichotomies related to the idealisation of informal care itself. I will start the discussion with an exploration of the specific meaning and connotations of informal care, in particular the construction and conceptualisation of care as 'being there for someone', rather than as a summary of particular tasks and/or skills. This traditionally unpaid, informal and privately based form of care entails an aversion to certain processes of institutionalisation, professionalisation and instrumentality. This will set the scene for understanding the moral condemnation of the commodification and professionalisation of care and related negative sentiments. This discursive process again relies on the creation of certain dichotomies between financial exchange and markets on the one hand and intimacy and close relations on the other. I will specifically draw on Viviane Zelizer's (2005) and Susan Gal's (2004) work to analyse the different aspects and consequences of this dichotomous understanding. However, as I have argued repeatedly, the ideal–typically constructed dichotomies are bridged and blurred, both empirically, and, as a consequence, discursively. I will thus focus on the relationship between care as idealisation and imagination and the conception of citizenship and rights. The summarising conclusion will try to answer the question of the degree to which the marketisation of care challenges or contradicts the current meaning of care and whether reconciliation is possible or even desirable, or, in other words:

> Can work be done for pay, and still be loving?
>
> (Nelson and England, 2002: 1)

The meaning of (informal) care

Developments in the context of care, such as the increasing professionalisation in the home care sector, the diversification of services, the introduction of cash-for-care schemes and the employment of live-in care workers, have challenged the boundaries of paid and unpaid, formal and informal care work (see Ungerson, 1997; Weicht, 2010). It has been argued that important differences in the nature of care arise according to whether it is provided in the public, private or non-profit sector and whether the spatial context is a private home or an institution (Held, 2002a; Timonen and Doyle, 2007; Zadoroznyi, 2009). As I discuss below, the idealised version of caring is strongly related to the presence and availability of a caring person. In particular, informal care entails meanings, associations and imaginations that reach far beyond the technical execution of certain tasks and services. Related to earlier discussions on the meaning of 'pure' relationships and a nostalgic imagination of community, the carer is thought of as being there for someone close to him/her, without asking for any (financial) remuneration. This holistic understanding of care also includes the imagination of a certain type of person who represents the characteristics of a carer. In a focus group discussion, Will expresses the fear that, because of current developments in society, people are losing this caring identity and the ability to be caring:

> Will: [It is] natural [. . .] to care for our elderly. But in-between we get greedy and selfish. And other elements of man come into play [. . .]. And so this caring element doesn't seem to be quite the same as

> I can't respond anymore [...] that somebody is just there. This for me would be ideal care, where you just don't have the feeling of being without anybody, being dead already.[1]

Tim clearly distinguishes between aspects of care work (or 'caring for') that can be delivered by profession carers equally well and aspects of care that can be described as *minding* or *being there for someone*, which he finds should be provided by people close to him. However, this separation of aspects that constitute care becomes more difficult in the context of the private sphere. Many people experience and imagine a unity of the different practical caring tasks and refer to the more holistic idea of being there for someone. Since the construction of the home is closely related to a particular conceptualisation of close relationships (see Chapter 3), care at home cannot easily be split up into specific tasks but requires the totality of the person who is the carer, as Adam in a different discussion emphasises:

> Adam: Yes, but at home [...] care and minding go hand in hand, that's blurred then. Especially if you are a family member yourself who cares for someone older. [...] You cannot really separate that then.

This suggested inseparability of caring tasks has important consequences for paid care. Lynch (2007), in her distinction between three forms of care labour, highlights that due to the historical arrangements for care, love labour, which describes the most intimate form of care, is not commodifiable as it is 'emotionally agaped work' with the principal goal being the well-being of the other.[2] Love labour is constructed as being clearly distinct from other forms of care work, being based on strong mutuality and 'the world of primary, intimate relations where there is strong attachment, interdependence, depth of engagement and intensity' (Lynch, 2007: 555). If care is based on the idealised notion of *being there for each other*, then emotional relations form a crucial part of the execution of care itself:

> The emotional work involved in loving another person is not readily transferred to a paid other by arrangement; neither can it be exchanged. To attempt to pay someone to do a love labour task [...] is to undermine the premise of care and mutuality that is at the heart of intimacy and friendship.
>
> (Lynch, 2007: 565)

Consequentially, the actors within this form of care relationship, Lynch argues, cannot be replaced by professionals or other paid carers. Feelings, such as love and devotion, shape the relationship between people and thus the care experience. This means that paid care inevitably falls short in substituting for traditional arrangements since while certain care tasks might be commodifiable, love labour is not. Lynch argues further:

> What makes commodification of care work problematic is the attempt to commodify the non-commodifiable dimensions of it. Mutuality, commitment and feelings for others [...] cannot be provided for hire as they are voluntary in nature. The love that produces a sense of support, solidarity and well-being in others is generally based on intentions and feelings for others that cannot be commodified as it is not possible to secure the quality of a relationship on a paid basis.
>
> (Lynch, 2007: 565–6)

Lynch distinguishes very strongly between love and the rational aspects of care work. An important aspect of her account is, however, that people need to be able to make a choice to commit themselves for the sake of the relationship and not for payment. In that sense, *being there* becomes *wanting to be there*. The idealised care relationship is thus characterised by both availability and a willingness to care. This imagined relationship needs to be understood as an ideal type that might look different in reality; the values, however, remain, penetrating the meaning and understanding of experiences of care. Marion, in the following extract, negotiates the ideal type's values with the realities people in caring situations face:

> Marion: It can go either way – the carer may be sufficiently fond of the recipient for their relationship to remain warm and loving, something which can't really be achieved when the carer is an employed stranger. On the other hand, perhaps a balance between the two is the best we can hope for.

The pricelessness of care

I have argued so far that the ideal-type construction of informal care emphasises values that clash with financial remuneration; in other words, the nature of caring work, as affection, love and emotional labour (Hochschild, 2003b), prevents it from being seen as commodifiable.

Caring in this definition is not motivated by instrumental factors but by concern for a specific other, an idea that strongly reflects an ethics of care. The commodification of care would thus mean the commodification of feelings (Hochschild, 2003a), which is seen as contradicting the idea of care itself. John's comment points to the intrinsic versus the emotional rewards of caring:

> John: People get so much out of it, not financially but emotionally, in terms of feelings. Apart from official recognition.

The emphasis on the pricelessness of care is not restricted to informal carers only. In Chapter 2, I discussed the different, sometimes contradicting, demands that professional carers face. Care workers find themselves confronted by demands, on the one hand, rationally to apply knowledge and skills and, on the other hand, to offer loving, devoted care. This tension consequentially translates into questions of remuneration. The following account, about the experience of working in care, also makes some reference to something other than usual working relations. Denise describes her work as mainly *making them feel comfortable and happy*, and the reward is a relationship of gratitude:

> Denise: I cared for people in a home, in a nursing home, I was only, just an ordinary – uh – dogsbody [laughs], care assistant, you know. Drinks, getting to bed, wash them [...]. For just 10 months, not for very long. But it really opened my eyes. And I wanted to carry on because I liked looking after them, because they were so grateful, didn't matter what you did, they were so [grateful], thank you, thank you. So grateful, even the smallest things [...] and that was the greatest reward, of making them feel comfortable and happy.

In public discourse, people often express frustration over the tensions between the nature of care as a loving, dedicated emotion and the requirements of rationality, bureaucracy and administration. In the following extract, the contradiction between care and bureaucracy becomes clear. Walter sketches out this distinction by referring to humans, *not a machine*, who should not be forced to think about money issues in the light of care for some elderly family member:

> Walter: The human is indeed a human being and not a machine or something like that. [...] Maybe one should be able to say yes, he needs so and so many hours, without thinking about the money.

[...] When I say, ok, if you need that, you get it. Whatever it costs. (...) The state has to pay for it. That's it.

This distinction between a *human being* and a *machine* again points to the association of real care with nature, and not instrumentality, as discussed earlier. Partly in contrast to earlier descriptions of the price-lessness of care, Walter's comments point to a gift of care for the care receiver. In his account, the state should secure the availability of any necessary care free of charge. While this idea could also refer to the provision of professional care, the discursive link between not having to pay for care and the emphasis on being human is reminiscent of the general construction of informal care. In general, however, this pricelessness is discursively rooted in the construction of informal care, offered within dedicated, loving relationships. The carer in this situation is constructed as offering a gift not only to the elderly person but also to society. The following newspaper extract demonstrates how the idea of heroes who give their lives for others is used politically in the public arena:

> Chancellor Gordon Brown unveiled the extra cash as he praised the unsung heroes of British society who dedicate their lives to looking after loved ones without being paid a penny [...] hidden heroes who keep families together.
>
> (*Daily Mail*, 21/02/2007)

Praising unpaid carers as morally superior and presenting those who care as heroes and role models in an otherwise selfish, materialist and cold society leaves many people without a choice. Caring is constructed as being outside normal citizenship and carers are affected in any choice they make simply because the discourse around care presents it as morally superior. However, this moral uplift goes hand in hand with the economic and social pressures faced by those who do care informally, as there is less time and space available to care for each other (Hochschild, 2003a):

> Eloise: So, [there is] really a lot of admiration, and personally, I also always find it really admirable if someone is taking that on at home. But, on the other hand, there's no money for it. [...] You can't live off admiration.

Emphasising the unpaid dedication of 'heroes' who are a role model for society has strongly gendered connotations. Held (2002a), for example,

informal and formal care arrangements, the two worlds are clearly separated. Zadoroznyi's (2009) definition of two forms of care reveals the most important distinctions. It thus becomes clear that underlying both informal, family-oriented and formal, professional care are moral constructions and ideologies:

> Most notably, informal care is *diffuse* (that is, unspecified), *based on feelings* (which might be anything from obligation to love); it is provided by 'identifiable kin and friends' on the basis of a generally *ascribed* relationship with the person being cared for; and is oriented to a *particular* person with whom there is a relationship and affective ties. In contrast, the logic of (formal) care provided by professionals is based on *functional specificity, achievement, universalism* and being *affectively neutral.*
>
> (Zadoroznyi, 2009: 271)

Informal care can thus be summarised as the presence (*being there*) of someone specific, within concrete, close relationships motivated by love and dedication. Importantly, these associations function through the creation of a dichotomous opposite: professional, formal care, motivated by rationality, functionality and financial remuneration. Discursively, care defined as being there for someone is contrasted with two fields of social engagement: work and politics.

Care vs. work

Distinguishing care from general notions of work means to emphasise the former's roots within traditional relationship practices. But how is care interpreted and constructed if it is delivered in formal, professional work contexts? Does the commodification of care change the nature of what care means to people? Kendrick and Robinson (2002) emphasise care's (and nursing's) nature as 'acts of loving', and Laabs (2008) points to religiously based roots of nursing, arguing that morality should bind strangers together 'as moral friends'. While work and employment are often identified with the (masculine) realm of 'hard' values, such as reason and justice, care relates to conceptions of the (feminine) realm of nature and natural emotions (Held, 1990; Hughes et al., 2005).[3] Held argues that due to the naturalisation of the separation of the two spheres, this dichotomy appears normal and essential. Similarly, Zadoroznyi (2009) points out that a normative view remains that ideal care emerges naturally from within family in contrast to a decision to offer it as a paid-for service. As described in Chapter 3, the construction of home as the realm of care plays an important role in situating care in opposition to institutionalisation and the instrumental, formal provision of services. Here, I will briefly discuss the consequences of the idea of institutionalisation for the meaning of the care work carried out in, for example, care homes. Guberman et al. (1992: 613) describe research, arguing that institutions are seen

> as being cold, rigid, normalizing places where feelings of love and self-sacrifice are totally absent. [People] were convinced that the care receiver could never get the same care there as in their family.

The fact that care in institutions is provided by paid employees challenges care's foundation on love, devotion and affect, almost by definition, as Walter remarks:

> Walter: There [in care homes] what is missing, [...] is idealism, of course this is only an employee there. You must not forget that!

However, the impossibility of the idealised notion of care is not necessarily related to care workers' willingness; rather, the reality of an institution being an unsuitable place for care restricts care workers in their devotion:

> Paul: It's the individual willingness and effort of those who work there. Whereas I don't want to say anything bad about them, that they don't want to, but that they are not in a position to do it.
> Ingrid: Yes, they try anyhow, as much as possible.

Professional carers are not seen as 'worse' people than those caring at home, nor are they criticised for delivering care in institutional settings; rather, the institution is, by definition (and additionally due to economic pressures), a realm in which intimate, loving care is not possible. In the next quote, Bea exemplifies the frustrations arising from the contradiction between an awareness of what care means and entails and how the institution is arranged and organised:

> Bea: The other thing is that when you're talking about carers, I know it's years ago, mid-1990s, [...] there'd be two carers for 22 people. [...] They'd all got to be bathed or washed, put to bed and given their nightly pills and they were lonely, it's simply because however kind you felt, you had to share. [...] They would say, share and talk

to them, of course they were full of what they wanted to say to you. And you could only stop and listen for a few minutes to that person.

In Chapter 3, I have, furthermore, argued that one's own 'home' is constructed in opposition to the outside world. I have stated that the traditional, bourgeois, middle-class ideal of home must also be seen as an antipode to the capitalist world of work, employment and markets. The home as an expression of care can be seen in a moral context in which a life in opposition to (labour) market forces can be lived. It is in care, and in particular care in the private home, where solidarity, self-lessness, family and community are seen to manifest themselves. This points to the contradiction described above, since one's own home is the quintessential realm of love, intimacy and care, and the institution must necessarily fall short against that ideology. As the institution is seen as the quintessentially uncaring space, people working and living in them are confronted with a discourse that defines their own situation in these terms. Obviously, this aversion to institutionalisation also has consequences for paid carers in people's own home. Martin-Matthews (2007) refers to an inherent contradiction in that people need to make sense of the 'stranger' in private places, referring to bridging the boundaries between the workplace of the carer and the home of the care receiver (2007: 233). Institutionalisation and professional care arrangements in people's own houses are seen as representing a market-driven alternative to informal care. Professionalisation and paid care are thus not discussed in relation to the quality of care but, to a large extent, as the ideological opposite of informal care. In the following focus group extract, the questions arose as to whether carers are seen as heroes and whether they should be paid for what they do. Betty's argument that payment decreases the value of care raises important questions for those professionally involved in care work:

> Betty: Most carers are happy to do the job and not get [anything].
> Nathan: Yeah. Most carers don't think of themselves as being heroes.
> [...]
> Betty: You can't put a price on it. If you're caring and you want to care for someone, putting a price on it begins devaluing it.

The question of caring as a gift has been discussed above. Importantly, payment is seen here as changing the care relationship and care itself. Real care, it seems, is not something that can be provided in exchange for money. The following extract from a different group takes up the theme of professional carers and refers to a contradiction between payment, professionalisation and dedication, here referred to as vocation. Interestingly, Pamela emphasises that vocation is not everything but that professional training is equally important. Her reference to Africa, though, already suggests that carers are seen as being a particular type of person:

> Larry: But, like nursing in years gone by, [...] to a certain extent it was a vocational profession. Someone wanted to go and look after others. Humanly help people that were ill.
>
> Pamela: And the best carers do have a vocation, [...] you can see it very clearly in the carers in my mum's home, that the best carers are the people who have come from Africa, basically, who are trained nurses in their own country [...] who have [...] much higher levels of training.

There is an inherent tension between categories of training associated with quality of care and vocation and having a caring identity. This tension lies at the roots of professional care and its differentiation from informal family-based forms of care. So, the question arises as to whether some forms of care work can be commodified while others cannot. To what extent is professionalisation a hindrance to loving care and, on the other hand, loving care a contradiction of commodified care? Are the two ideal categories mutually exclusive? Lewis (2007) doubts whether all care can be commodified, arguing that care is not only a task but also an emotion, and that unpaid care by friends and relatives can never be fully substituted by commodified versions of care. It must be noted here that this bipolar construction does not only result from a reference to *'professional carers, whose commitment may be questionable'*, as the *Daily Mirror* describes it (06/04/2007), but is generally related to an ideal of care based within a realm of dedication, emotion and affection. By constructing care in opposition to work, with an emphasis on natural values of love, affection and dedication, and in contrast to materialistic goals and motivation, care for the elderly is portrayed as a model of ideal, selfless and committed behaviour. Good behaviour is given out of love and selflessness and can therefore not be included in the logic of the market and payment for labour. In contrast to work, care is discursively related to particular kinds of relationships (see Chapter 2). Ungerson (2004) demonstrates that different forms of funding have different impacts on the nature of care relationships. In response to Helma's account about a professional carer and her own experiences as a voluntary visitor to elderly people, Uta emphasises that payment and employment do create specific working relationships, which are distinct from unpaid caring relationships:

> Uta: And I do think that's a big difference [...] such a carer who [...] – first of all she gets paid [...] which means she is my employee, that's also how it'll be seen by many. Not everyone will be so nice and lovely like the one in [the village] is. [...]
> Helma: That's of course true. [...]
> Uta: And what you're doing, that's a voluntary social service, which is unpaid, which means I arrive there, chat with her and I don't get an advantage from it. No use. While a carer is paid for.

It is argued that if financial remuneration is involved, the relationship fundamentally changes. Consequentially, the voluntary provision of care falls on the side of informal, family-related care and is constructed in opposition to professional, paid labour. The classical distinction between (informal) care and work does not mean, however, that informal carers are always seen as selflessly wanting to offer their devotion and services. While Theodora in the following short statement confirms that informal care might be given out of particular feelings of obligation, or even coercion, the discursive contrast with paid labour remains:

> Theodora: Informal care is done through emotional coercion and, dare you ask, for recompense. Paid labour is like any financial contract; you do a job and expect a living wage.

Care vs. politics

A dichotomy that builds the basis for opposition to payment and the sphere of formal work is obviously related to the classical split between the public and the private (see, for example, Chapter 3). Even though the associations with masculine and feminine realms have changed over time, distancing from care and thus from the private sphere has always been an important element of the construction of the opposite, the masculine, public and political realm (Leonard and Tronto, 2007). Likewise, in discourse, care is positioned as an ideal (which is associated with the private) that should be protected from political competition. Official (party) politics in this context is seen as inherently opposed to 'real' care. People continuously criticise the involvement of public agencies to control and check on carers, and caring facilities as these institutions intervene with the real nature of care. In the negotiations between care and politics, some interesting diverse discursive patterns emerge: on the one hand, people want politicians and the state to act by providing support and intervening when care provision falls below a certain standard; on the other hand, they reject making care a political issue (see also the discussion on rights and citizenship later in this chapter). Politics therefore fulfils two functions in the discourse on care: firstly, people identify it with official representation, negotiations, business and economic decision-making. Secondly, people see politics as a substitute for 'the state' or 'the society'. In this sense, people express the wish that 'politics' should enable and foster care. Care itself, in particular in its edifice of 'being there for someone', is constructed as non-political or apolitical, that is, as an issue that should not be a topic of political argumentation, campaigning or legislation. In this context, Aldridge (1994) points out that constructing issues as apolitical often implies a certain moral relevance that cannot or must not be contested. This moral value becomes apparent in the following editorial commentary in the *Daily Mail*, which, in reaction to Gordon Brown's contribution to the newspaper (see above), writes:

> Some issues should be above party politics. The treatment of carers is one of them. They are the cement which holds the nation together, selflessly giving up their lives for the sake of those they love. We applaud Gordon Brown for recognising their worth.
>
> (*Daily Mail*, 21/02/2007)

Political competition is thus constructed as belonging to a sphere of rationalist, materialist decision-making. This is again contrasted with an ideal of care and community that opposes the world of work, markets, politics and de-personalised relating. Using this particular newspaper extract in focus group discussions, I have prompted people to think about the relationship between politics and caring. Again, a real distinction is made between political confrontation and competition and the values of care. Below I quote from a discussion in more detail as it highlights well the ideas and themes that underlie the aversion to politics in the world of care. The extract shows the construction of two distinct spheres: the world of party politics, its relation to competition, profits and rational behaviour and the world of care, which is based on emotions and feelings, rather than on negotiations and conflict.

Patricia: I think it should be above party politics
John: I agree with that.
Patricia: Never going to get it above party politics, unfortunately.
Nathan: It becomes a weapon [...] for the parties. But I think it's true that [carers] are the cement which holds the nation together. Because they do, carers, we do save this country a lot of money. [...] So, in a financial sense, it cements it together but relationally it does too.
I: And why should it be above party politics? What do you think?
John: Party politics for me is, it seems that you can get into a quagmire, there can be a lot of [...] one-upmanship, gamesmanship and all these things can come into play and actually, the thing about that, it can cloud the issues, and once you get [...] issues regarding care involved in that cloudiness, you're not gonna get a clear picture, you're not gonna get a clear vision out of that. And some issues need to be kept out of party politics [...]
Peter: Who benefits from party politics? It's not the person who needs the care. [...] It is the person who engages in politics. Are we going to be caring for people or are we gonna talk about it? [...]
Nathan: I think, what I would say is that everybody should be a carer. [...] Not making caring a sort of political football. There should be a general sense of humanity, that we look for the best, for each other. And that we work to that end. Rather than it just being, I mean it's something that parties have to talk about, they're gonna have to come to conclusions, [...] the spending of money etc., there's gonna be disagreement. But when it just becomes party politics, for the sake of a political football, then that's a different [story].

The ambivalent position of politics becomes evident in this extract. The discussants clearly reject (party) politics as being opposed to care. The imagination of real care is described by Nathan's idea that it reflects *a general sense of humanity*. When Nathan argues that *everybody should be a carer*, it again favours the direct, natural engagement of people over political, abstract decision-making processes. At the end of the extract, however, Nathan acknowledges the need for politics, in the sense that *they have to come to conclusions*. The paradox of simultaneously rejecting politics in the context of care while formulating demands from the state describes one element of the empirical and theoretical blurring or even attempt at reconciliation of the two opposites. In shaping people's understanding and therefore the meaning of care, however, the dichotomies retain relevance and significance.

Beyond the dichotomy

This chapter has thus far focused on the ideological construction of two separate spheres of informal care on the one hand and commodified care work on the other. I have also pointed out that those providing care are confronted with this moral and ideological construction. The construction of the person of the carer is embedded in the two-world dichotomy (which spreads into different spheres as discussed above). Empirically, however, an understanding of care as being clearly divided between informal, family-based care at home and commodified, professional care in institutions is misleading and does not reflect the reality of hybrids of love and instrumentality (Ungerson, 2000) and contract and affect (Glucksmann and Lyon, 2006). Examples of intermediary arrangements, such as non-profit services that are not necessarily governed by market principles (Held, 2002a) or voluntary schemes that focus on the 'altruistic and idealistic motives of volunteers' (Glucksmann and Lyon, 2006: 6.2), are proof of the blurring of boundaries of this dichotomy. In Chapter 3, I used Gal's (2004) concept of 'fractal distinction', which emphasises the ideological, discursive quality of the creation of separate realms of public and private. Importantly, Gal (2004) argues, this discursive pattern of dichotomisation is replicated within the opposites as well. The distinction between informal and professional care does not complete the process of construction; rather, the division is reproduced within the sphere of professional care. Similarly, Lyon and Glucksmann (2008: 114) argue that 'a simple dualism [...] cannot readily distinguish between different kinds of commodity or non-commodity relations'. Clearly, the distinction between real care and work does not strictly follow the trajectory of the division between paid and unpaid or formal and informal work. As a consequence, some people in professional care do work, whereas others really care. Similarly, commodification and professionalisation, while in principle being constructed as oppositions to the ideal of care, can entail levels and hints of the ideal. Importantly, however, all empirical blurring of the boundaries of the dichotomy has to deal with the moral associations inscribed in the different poles.

Several empirical arrangements try to combine financial transactions with intimate care. It can be argued that due to economic and social developments, a combination of these spheres seems desirable. In fact, markets and state arrangements do play an important role in many societies' organisation of care (Ungerson, 2005). The logic of the market has challenged the moral understanding of care (Glucksmann and Lyon, 2006), and new forms and mixes, 'which transcend the public/private, market/non-market and paid/unpaid distinctions, as well as the love/money/duty nexus' (Glucksmann and Lyon, 2006: 7.1), have emerged. Do these changes mean that the meaning of care is altered as well, according to different contexts and in diverse socio-economic modes and locations (Glucksmann and Lyon, 2006; Lyon and Glucksmann, 2008)? Glucksmann's (2005) total social organisation of labour approach shows that there is interconnectedness across boundaries between paid and unpaid work, market and non-market, formal and informal sectors. For example, intimate care by professional strangers (Karner, 1998; Ungerson, 1999) is happening in people's own houses, and the employment of migrant carers has shown how marketisation of care can be combined with informal, illicit arrangements in which care workers can even be 'adopted as fictive kin' (Weicht, 2010). Barker (2002) furthermore identifies 'unpaid, nonprofessional nonkin caregivers' as bridging separate spheres which are underpinned by a moral construction. Zelizer (2005) also argues that we are constantly mixing relational intimacy with economic transitions, but that the ideology of two separate spheres persists. This causes many difficulties for those employed in sectors which Hochschild (2003b) calls 'marketized private life':

> Each realm has its own kind of feeling rules. If those in the realm of work follow the feeling rules of a company, and those at home rely on the feeling rules of kin, those in marketized domestic life draw on complex mixes of *both* work and family cultures.
>
> (Hochschild, 2003b: 204)

If financial exchange penetrates the private sphere of care, does this fundamentally challenge the wishes and desires associated with care? As I have established earlier, care is largely being idealised as 'being there for someone', as a willingness and ability to be present and available, both physically and emotionally. But how far is this ideal questioned when certain services are sold and bought rather than provided out of love and dedication? Does the commodification of care contradict the imagined caring mindset?

Holistic provision of care

Care is based on the presence and availability of 'a carer' (Degiuli, 2007), referring to a particular identity constructed in the discourse on care. But how is this identity of the carer shaped in relation to a discourse emphasising values such as empathy, love and affection? And how can professional carers be described and understood in this context? In its discursive construction, a continuation of the dichotomy presented above can be found in the split between *doing* and *being*, whereas the latter signifies the sought-after identity of a natural carer and the former reflects the tasks performed by a care worker. The dichotomy between real care and doing a job as an employee is not only a difference in arrangements (formal and informal) but also in mindset, attitude and character:

> Fran: But then that varies, some carers do really care, and love their job, and others are just doing their job.

The dichotomous construction is thus reproduced in the descriptions of professional care workers. Mary in another discussion similarly argues that both professional and family carers really can be a caring person, but *in a different way* both really can be there for someone:

> Mary: I personally would say if the person's doing well, if the person has everything, then she's well looked after or cared for. And I believe that both sides can do that well. And, of course, in a different way.

Real caring, albeit in different ways, is thus to some extent possible in both the formal and informal sectors. Commitment to and love for people are themes that distinguish real *care* from the performance of *work*. The question arises as to what constitutes a caring identity for professionals and what aspects are shared between formal and informal carers. Adam, in the following extract, distinguishes between some people who are seen as only doing their job and others who love the person in need of care:

> Adam: You also need to have a certain vocation, or love for it, because if you'd rather become a builder and you are then supposed to care, you couldn't do that of course. There are probably also always people in this area who do not 100 per cent fit it, [...] who just do

that. You can see it in hospitals, they do their job, but they don't do anything more than that, and they just do it monotonously, like on a conveyor belt.

Barbara: So it needs a lot of idealism to do a job like that, that's your statement, isn't it?

Adam: Yes.

Barbara: yes, and also love for the older person in need of care.

Love, dedication and commitment form the idealised virtues of a carer, regardless as to whether he or she is caring for a relative or for a stranger for payment. So, is real care then possible in institutions if the right people are present? What enables or hinders real care within a professional, institutional setting? Nathan, in the following discussion, emphasises that, in the context of hospitals, nurses would like to care, but the institution, however, prevents them from doing so. In other words, the institution is organised as a workplace, not as a realm of care:

John: And I'm sure that if you are in an environment where you do have these dedicated staff who really care and [...] are dedicated to what they do, then I think...

Nathan: [...] Part of the problem in our care system now is that nurses who went into nursing to nurse are no longer permitted to nurse. That they find themselves in situations where they are managing wards and it's become very management-structured. It's become very much [...] time managing [...] So if you're going to A&E, if you're there for longer than four hours the hospital gets a fine, so you're pushed out to a ward somewhere, or you get, even worse, put in an ambulance and driven around to other hospitals just so that that the hospital can actually hit its target and [...] not get fined. That isn't care to me. [...] But nurses like this [points to a picture] are wonderful. It is a vocation, like Peter said, [...] it is a vocation, but nurses are pleading in this country to be able to nurse. [...] Rather than meeting targets, rather than just being in situations where they are so short-staffed. [...] And it's not the fault of the nurses. The desire of the nurses that went in to nurse is that they do nurse. But they find themselves under so much pressure [...]. And it is finance-driven again. [...] You cannot set a budget to care in this country.

The dichotomy of loving, dedicated care and the performance of work continues to play a crucial role, even though people acknowledge values and virtues related to care in both informal carers and professional care workers. In the latter context, however, an inherent tension arises. The institution and work circumstances are constructed as unloving, cold and rigid places (see Chapter 3) in which individuals cannot perform real, loving care. Both care receivers and caregivers are thus confronted with and have to overcome a logic of professional, institutional, bureaucratic arrangements. The relationships within professional care are thus complex and ambiguous. Performing a role that is based on a 'contradiction between command and obedience on the one hand, and sensitivity to feeling on the other' (Ungerson, 1999: 586) challenges the imagined ideal of the identity of caring for and caring about. Elsewhere, Ungerson (2005) distinguishes between different emotional variations of relationships, labelling them cold, cool, warm and hot relationships, according to the closeness to family-based idealised care relationships. Ungerson argues that some commodified relationships can even represent hot relationships, as is the case of migrant carers in people's homes (for a discussion on migrant care workers in Austria, see chapters 2 and 3):

> Such relationships are unprofessionalized and unregulated in exactly the same way as non-commodified informal care relationships are [...]. But both sides in these relationships are vulnerable to forms of exploitation and even abuse particularly since the relationships are acted out behind closed doors within the domestic domain. Given the core vulnerability of frail old age on the one hand, and illegal immigration on the other, combined with spatial proximity, very low wages, and twenty-four hour availability, it is not surprising that these relationships are full of feelings, not all of them healthy or likely to underwrite reasonable quality care.
>
> (Ungerson, 2005: 202)

In other words, these arrangements do not represent work in its discursive construction but refer to an idealised notion of relationships. In this quote, Ungerson also points to the potential vulnerability to exploitation involved in all caring arrangements, particularly in those that reflect traditional *caring* relationships. Macdonald and Merrill (2002: 67–8) argue that altruism, empathy and emotional involvement are inherent to the nature of care, but that care workers are 'denied recognition of, and compensation for, this investment of self'. They argue that because care is constructed as 'nonwork' and (female) carers as 'nonworkers', people miss out on a full partnership between them and others (Macdonald and Merrill, 2002: 75).

In describing the nature of care, I have argued that care is constructed in a way that makes it impossible or undesirable to distinguish between different tasks; rather, the identity sketched out is one of simply *being a carer*. In other words, someone is caring if he or she is there and is involved with a particular, concrete relationship. Ungerson (1999: 598) describes the nature of the work of personal assistants (whose occupation is to some extent similarly constructed to those of informal carers) as 'essentially *unorganised* and *particular*'. The main good given by the carer is time, in particular being 'flexibly available, normally for very long periods' (Ungerson, 2005: 193). This element of flexibility is a core feature, as the carer as a person is constructed as someone who is 'compassionate, emphatic, merciful and selfless' (Winch, 2006: 14). And this construction facilitates persuasive pressure on people to work beyond their contract hours and conditions (Ungerson, 2000). Ungerson (2000) points out that social, professional care, seen 'as a set of tasks', can easily be commodified; the nature of care, however, leads to a situation in which even paid workers are constructed as behaving similarly to informal carers in introducing a feeling of being involved 'in the provision of total care' (2000: 630). I have demonstrated this process in the discussion on the employment of migrant carers in Austrian families, who are paid while 'becoming family members' (see chapters 2 and 3). People want someone who cares, someone who is there for them. This must be distinguished from someone who does care work, who is performing certain tasks. In the latter case, this can be bought; in the former case, however, this is not an optimal option, as the following short extract exemplifies:

> Caroline: That caring does not only mean [...] I do everything [...] but that caring also means I'm simply there for you and try to keep your dignity as well. [...] And this, relatives can often do better, I think, than trained nurses.

It is important to understand that this aspect of being there for someone cannot be split into separate tasks. Carers in an ideal situation *do* care and are not following certain procedures. In the following account, real care is described as being there, 'looking after the whole person':

> Nathan: The other thing about care is, [...] we think in terms of care [...] for the elderly, as being the individual person. But someone who's coming in and looking after the individual person doesn't actually cut the grass, or cut the hedges. Doesn't actually look after

the whole person. So for my [relative], I would come and at certain times he was looking out of the window and seeing the sunshine and seeing the hedges growing around the area. He would say to me: 'Nathan, I can't see the people walking by, over the hedge. Can you cut my hedge please'? Now, a carer would never get a pair of shears [...] and say: part of the care for you [...] is that you want to see your neighbours going by, you want to be able to wave, you want to be able to see, hear the children going by when they go to school, so you want your hedge at a certain level. [...]

Peter: Do carers do something beyond their remit, then [...]

Nathan: And then cutting the hedge was never in the carer's remit. She was asked, well, she should have come in and helped him wash, get him tidy, get him clean, get him dressed, prepare his breakfast, tidy up where he lived.

This description points to the nostalgia discussed above (Chapter 4), but here I want to emphasise the construction of a caring identity that seems to get lost. The inability to *be* a carer is often explained by references to broader market-related social and economic developments in society. As indicated in Chapter 4, real care is often constructed as something from the past, in which people would not have restricted their caring to certain tasks but *would have gone the extra mile* in their spare time. The question arises as to whether this represents a particular identity that is necessary in order to *be* a caring person, rather than *doing* care work. Is there a general perception of carers as specific types of individuals (Lloyd, 2006)? A perspective drawing on Virtue Ethics in this context would stress the importance of certain characteristics needed by someone wanting to be a caring person (Groenhout, 1998). The following discussion shows how these virtues are also constructed in the context of formally employed carers:

Morgan: That's what's lacking a lot, I mean, I have to say [...], I've worked in various hospices, there is just such a culture in there that nothing was too much trouble. The people who work in there, it's supposed to be a burn-out period but they all went beyond that because they loved it and they found it rewarding and so they would do whatever was asked of them. If you look at nurses, you walk into [any] hospital, you're not gonna find nurses like that.

Larry: That's a different kind of person.

Morgan: Absolutely. [...] they were totally committed to what they were doing and understood what was required of them and were able to give more than was asked, I think.

Will: Oh yes and I have just seen a [...] hospice which is a day-care hospice [...]. But I mean [...] look at the volunteers, it wouldn't exist, we don't get any money [...] from the government [...] And the volunteers are absolutely wonderful people, it just [comes] out of them [...] the care that they have.
Morgan: It's a privilege to be around them.

Commitment and dedication as values are emphasised (the volunteer personifies these virtues), but there is also a strong notion of the character of caring people, both paid and unpaid; they are a *different kind of person*. I emphasised earlier that many associations with care, the private sphere and a particular constructed identity bear important gendered connotations. Not only have women been continuously associated with many of these values and virtues, but the underlying dichotomies themselves represent gender-linked discursive categories. What can clearly be seen in the following extract are the gendered aspects of the (informal) carer's identity, which shows many parallels between the identity constructed and the traditional, stereotypical female identity. A distinction is made between the care business and *touchy, feely hands on care*, the latter representing an aspect associated with women:

Larry: it's the women who do the caring. I've seen that. It's not a men thing. Yes there are certain men in the care business as such, but when it comes to the touchy, feely, hands on [...] And they don't care whether they're looking after men or women, it's the female of the species, [laughing] who in reality is the hand-on person [...] who will go in and clean up an incontinent person or something,
Pamela: That's right. [...]
Larry: And this is a more natural [...] thing, is it not?
Morgan: Yeah, but when I was up in [a city], they have a voluntary sector organisation up there and that provided voluntary services for people who were dying in their home. And some of those were men, and they were telling me stories, particularly when it was a chap on his own [...] who was dying, they've gone into the home, sat there, held his hand, until he died, and he'd gone to the funeral, befriended the family and all of this. And I thought they were absolutely amazing, [...] and there were quite a few of them, so there's nothing stopping them.

Again, it can be observed that even though it is acknowledged that both men and women are able to and actually do care well, gendered dichotomies are still an important discursive feature. The caring identity, which resembles traditional constructions of women's role and identity, can thus be embodied by men and women, informal and formal carers alike. All of them, however, are confronted with a separation of their identity from the world of bureaucracy, markets and professionalisation. While the aversion to markets, institutions and bureaucracy is reproduced within the discussion on the professional care sector, care's general associations remain largely intact. On a societal level, this means that changes in the provision of care (including various formal arrangements) are understood as necessary and inevitable. Discursively, however, these developments require an adaptation to or recalibration of traditional dichotomies. The role of the state is particularly interesting in these attempts to renegotiate the boundaries of dichotomous constructions.

Rights and citizenship

As I mentioned earlier, the introduction of cash-for-care schemes has been one of the defining features of European welfare states' reactions to rising care demands. With the payment of cash benefits to fund individual care services, people's choices and autonomous citizenship rights were secured and increased (Rummery, 2009). Isabel Shutes (2012) argues that the attention to issues such as choice and control over care constitutes shifting social relations between care users, care workers, their families, employers and the state, and this is indicative of more general changes in the relationship between the individual and the (welfare) state. The organisation of how individuals relate to each other, and how the state relates to the individual is constitutive of the arrangement of citizenship and the role of the state in people's private lives (Kivisto and Faist, 2007; Lister, 1997). Arrangements of care need to be understood as a crucial part of citizenship itself (Knijn and Kremer, 1997). How a country defines its respective relations can be described by the concept of citizenship regimes, which include not only institutional arrangements and national rules but also problem definitions and assumptions about the conception of identities (Jenson and Sineau, 2003):

> A citizenship regime encodes within it a set of identities, of the 'national' as well as the 'model citizen', the 'second class citizen', and the non-citizen. It contributes to the definition of politics that organizes the boundaries of political debate and problem recognition in each jurisdiction. It encodes representations of the proper

and legitimate social relations among and within categories and the borders of public and 'private'.

(Jenson and Sineau, 2003: 9)

The construction of care takes place within an ambiguous realm outside traditional, masculine notions of citizenship on the one hand and a necessary societal task that forms part of the citizenship regime on the other. In the context of care, different actors embody different notions, and levels of citizenship, and dependent, elderly people are usually not regarded as being able fully to perform their rights and duties (Weicht, 2013). What does this role of care within the citizenship regime mean for the discursive construction and negotiation of dichotomies? Two aspects appear particularly interesting in this discussion. First, care's meaning as 'being there for someone' is translated into a request that care should not need to be asked for or demanded, but rather just organised for people who are in need. As Gita puts it, care at home, by the family, should simply be funded, without the need for forms, assessments or other evaluations and calculations:

> Gita: And that's why I say, that's of course no question at all, that at home, care at home was the best, right. [...] And I don't understand why care at home is not simply financed.

The imperative for society is to provide for a dignified life and the objectification of needs and wishes consequentially culminates in the idea of a right rather than benevolence. Different newspapers feature many commentaries in which the *right* to a dignified life is specified, as, for example, a healthcare professional and consultant in *Kurier* (10/02/2007) writes:

> I plead for a constitutional right, that everyone who lives and works in this country receives help from the state in their old age.

The state's role is constructed as one that should guarantee the provision of 'dignified care' in people's homes, but at the same time old age and care should not be a field for political controversy. What is important here is to draw attention to political implications that are often masked by a focus that too strongly emphasises the social and cultural factors of care. The second interesting aspect of the relationship between care's discursive construction and its place within conceptions of citizenship concerns the role of the state and society. Many contributions to both newspapers and focus group sessions declared that the state (and society) should look after its citizens. At the same time, however, some element of self-reliance was raised by the participants, usually in the sense that rich people should not spend all their money and later expect society to care for them. But if people have been worked hard, then society should provide for them later on. The state's role is crucial insofar as it is seen as being to secure non-profit arrangements for care, which still allow an informal, loving relation. In Austria, for example, this desire has been translated into a sentiment strongly favouring social insurance solutions, if necessary, over for-profit, market arrangements, which are seen as not caring by definition:

> Ida: It would have to be social insurance which is not in it for some private profits.

While the general construction of dichotomies suggests a negative sentiment towards politics, bureaucracy and public arrangements, the potential to avoid market-related care services facilitates a more diverse, ambiguous understanding of the role of the state. Importantly, while this observation blurs the boundaries of a dichotomous construction, the general aversion to markets in the field of care remains a persuasive discursive item. Care and citizenship thus form an interesting and challenging relationship in which the discursive construction situates care outside the traditional realm of citizenship, but in which an emphasis on citizenship rights allows one to reject the for-profit market logic of care. Paul Kershaw (2005) therefore suggests a redefinition of both care and citizenship and a necessary reconciliation within a new theoretical and political framework. Importantly, both receiving care and the possibility of providing care without economic and social repercussions need to be constituted as a social right of citizenship (see Knijn and Kremer, 1997; Lister et al., 2007; Orloff, 1993). Conceptually and ideologically, care itself, Kershaw argues, should be seen as a crucial, constitutive aspect of citizenship:

> the solution rests in embracing private time for caregiving as a constitutive element of full social membership. Since care work is critical for social reproduction, it is reasonable for the public to expect all citizens to make a minimum contribution to society's care needs as a condition of benefiting from the mutual advantages made possible by community collaboration.
>
> (Kershaw, 2005: 129)

Public definitions and arrangements of citizenship and rights thus directly contribute to the construction of the meaning of care by designing the frame in which public and private arrangements (and their intersections) take place. At the same time, the meaning of care shapes people's conception of citizenship and the role of the state in interfering in private matters on the one hand, and protection from the sphere of the markets on the other.

Conclusion

The intrusion of markets into the realm of care is both an empirical fact and a political and discursive challenge forcing us to rethink and reconsider current constructions of the meaning of (informal) care. In this chapter, I have described the nature of the construction of care as involving intimacy, feeling and love, and as an ideal of 'being there for someone'. This meaning of care involves strong discursive and emotional aversion to institutionalisation, professionalisation, instrumentality and politics. Care is idealised as a gift of one's physical and emotional presence, provided within concrete, close relationships. This idealisation is based on the construction of a dichotomy of care and work through which care is positioned in opposition to the world of markets, paid work and economic individualism. While this dichotomy underlies the discursively constructed meaning of care, the empirical provision of care services is often marked by a blurring and bridging of two opposites. In concluding this chapter, I want to discuss the discursive renegotiation of the dichotomy of care and work by returning to the main questions: Does the discursive construction of care in opposition to markets make the professionalisation and/or commodification of care impossible and/or desirable, or is a combination of intimacy and markets morally thinkable? What possibilities are there for social and political interventions in order to improve the situation of those in need of care and for those wanting or having to give or do the care?

Personal commitment, devotion and dedication have been described as crucial markers of the provision of real care. Whether professionals or informal carers perform care services, it has been shown that a particular identity of a caring person is sought. The discourse shows the imagination of and praise for a particular caring mindset that is necessary really to *be* a carer. Even in institutional settings, professional carers can be identified with those very characteristics. Additionally, care practices themselves can create and shape emotional proximity. Ungerson (2005: 189) highlights that as care relationships involve physical touching they have the potential to promote specific forms of intimacy between strangers. The correct attention to intimacy and closeness thus seems, to some extent, to be more important than the formal arrangements. In the context of the commodification of care, one can thus speak of the 'marketization of intimacy' (Ungerson, 1997: 363). Importantly, as I mentioned earlier, emotions, intimacy and financial remuneration are not necessarily mutually exclusive; in many cases, these sentiments might form the very content and objective of what is bought and sold (Hochschild, 2012).

Empirically and discursively intimate, loving care and professionalisation, marketisation and institutionalisation recurrently intersect. The discourses show, however, that these intersections require a renegotiation of the two-world paradigm. At the same time, however, the main building blocks of this opposition, the contradiction between intimacy and markets, resume their persuasive, discursive function in separating different logics of care provision and relating to each other. Using Gal's (2004) concept of 'fractal distinctions', I could show that the discursive split between intimacy and markets is, furthermore, reproduced in different areas of professional and informal care. Like Zelizer (2005), I claim that the construction of two hostile worlds needs to be challenged, reconciled and overcome. Furthermore, the ideological and discursive distinction between the idea of a morally good person and the economic, political sphere of transactions needs to be questioned. This would help to facilitate an understanding of the particularities of care and the difficulties for those who do this work. Zelizer (2005) argues further that the hostile worlds of sentiment and rationality have serious practical implications and divert from real solutions. Analysts of care (Lloyd, 2006) have pointed to the problem of a systematic distinction between paid and unpaid carers in society, with the political and economic system needing 'carers to be heroic and self-sacrificing' (Lloyd, 2006: 952). Harris (2002) identifies a remoralisation of discourses under New Labour in the United Kingdom, which led to a reproduction of the split between economic exchange and emotional intimacy. A rethinking of the nature of relational work (Zelizer, 2005) seems to be required, which includes an assessment of markets, their limits, promises and consequences (Karner, 2008). In order to avoid the exploitation and misrecognition of both paid and unpaid carers, new concepts of making sense of the worlds of intimacy and economic relations need to be devised. In Chapter 5, I argued that asymmetrical relationships need not be harmfully hierarchical (see also Nelson and England, 2002) but are normal aspects of human existence. At the same time, professionalism and employment do not necessarily mean non-attachment, or an anti-emotional approach. Thus, the question is not whether some commodification is better than others, or whether commodification and professionalisation are better in some parts than in others. What is needed is a reconsideration of the fractal distinctions (Gal, 2004) present in the discourses on care. If economic exchange and intimate, loving involvement are not understood as contradictions anymore, political intervention can create new arrangements for those caring (in a professional or informal capacity) and those being cared for (at home or in institutional settings), which live up to people's desire for someone providing a loving, intimate 'being there for someone' feeling, but which at the same time do not create vulnerability to exploitation for all those involved in caring relationships. I have also noted in this chapter that politics has to manoeuvre in some contradictory discursive realms. On the one hand, politics represents a world of markets and rational decision-making, which is rejected; on the other hand, however, political decisions are desired that will allow real care and offer protection from intrusion by the markets. I have argued that politics needs to be understood in a positive context as a substitute for state and society, as a concept that encompasses all levels of society. The paradox of the state's role is based on the tension between politics and markets, which are, on the one hand, seen as representing the same sphere; and on the other hand, politics is sought that offers an escape from the market. These tensions and ambivalence need to be taken seriously as they underlie the ideological and moral construction of care.

Recognition of the ideological nature of the split between care and the economic sphere is essential in order to create a more just, democratic and compassionate system of caring for each other (Zelizer, 2005: 303). This, however, also means an earnest and genuine appreciation of the motivations for the discursive construction. People's associations with and imaginations of ideal care need to be taken seriously in order to devise the successful commodification of care. An ethics of care based on 'an understanding of its intertwined values, such as those of sensitivity, empathy, responsiveness, and taking responsibility' could help to 'adequately judge where the boundaries of the market should be' (Held, 2002a: 31). Caring and concern for each other are values vital to the functioning of society and should be appreciated (Held, 2002a). Any process for the commodification of care needs to be aware of the significance and substance these values possess. Responsibility, dedication and empathy also form important aspects of a more caring form of citizenship. Nancy Fraser (2008) emphasises how, for disadvantaged or marginalised groups, besides redistribution and recognition, the possibility of participating in decision-making processes constitutes a crucial element of regaining justice. If equal participation of those most strongly affected by arrangements of care, people in need and those providing caring services, is an objective, then people's rejection of profit-driven market logic in the context of care needs to be taken seriously. The analysis of discursive construction might thus provide the basis for a form of democratic care (Tronto, 2013) in which the allocation of caring responsibilities is based on just and responsible attention to people's concerns. The commodified version of care needs to fit in a logic and desires associated with care in order to be a viable option:

> The question about whether the market can be caring depends upon how members of a society think about the market and its purposes, and whether the market is accommodated to fit with other institutions.
>
> (Tronto, 2013: 115)

The marketisation of care has obviously created many challenges to and pressures on the system of care and for those navigating within it. In various contexts, the logic of profit can and has seriously harmed the quality of the provision of care and hence the experiences of both carers and care receivers. However, this chapter has shown that the aversion to marketisation goes beyond a critique of the empirical consequences of this process. Rather, payment and markets generally symbolise an opposite to the imagined caring ideal. The answer to the question posed at the beginning of this chapter, whether work that is paid for can still be loving (Nelson and England, 2002), would be yes. Remuneration, professionalisation and institutionalisation do not by definition alter the meaning of care substantially, as long as the main associations of care are not challenged. Introducing and relying on a logic of profit, however, seems to contradict and damage the ideal image of 'being there for someone'.

Epilogue

The meaning of care, as I have argued throughout this book, is constructed through discursive practices, which are shaped by and based on concrete relationships between people. By analysing the discourse on care, I have tried to identify people's associations and assumptions about what care *means*. I have investigated how moral attitudes and moral concepts are constructed and reproduced in the context of care. It has been shown that people use this moral grammar to make sense of an important social practice, care for elderly people. Naturally, care is a very personal issue, a personal experience and a personal challenge for everyone. How care is arranged, experienced and imagined reflects people's personal biographies, ethics, relations and constitutions. However, in this book, I have attempted to show that while the meaning of care is fundamentally shaped within concrete social relations, the construction of those relations, and thus also care, needs to be understood as a fundamentally social issue. In his conception of the *Sociological Imagination*, first published in 1959, C. Wright Mills (2000) urges us to investigate personal experience in relation to and in the context of wider societal issues. Personal biographies inevitably intersect with social and cultural history to give meaning to practices and thus make them social issues. I have therefore focused on the social construction of the moral meaning of care, which influences individual decision-making processes and impacts on the general understanding of caring and being cared for. What I hope this study has achieved is to fill a gap that appears in much of the care-related literature, namely a focus on the relationship and interrelation between social practices (caring), social policy arrangements and ethical and moral constructions of society. Throughout the book, I have demonstrated that care does not fit into the economisation of society, nor does it fit neatly into an individualisation thesis. Rather, care reflects an ambivalent desire of people, which can be described as *being there for each other*. This construction has important consequences for all those involved in caring relationships (as carers and as those being cared for). Those involved in care cannot be characterised by the tasks that need to be done; rather, people's identities as *being the carer* and as someone in need of care are defined. Due to the construction of care as a moral practice, based on love, intimacy and *being there for each other*, people involved in those relationships are vulnerable to exploitation and face a marginalised position in society.

Focusing on core aspects that constitute the meaning of care, I have presented a discursive image which is created through dominant narratives, experiences and contributions. I have argued that the construction of family care is a representation of an imagined ideal, which can also be embodied by non-family members. In that sense, there is no straightforward assumption that family members are exclusively responsible for the provision of care for their elderly relatives. Chapter 2 has thus demonstrated the importance of aspects of closeness and relating within care. At the same time, 'natural' traits, attitudes and opinions do play an important role in people's understanding of 'the proper thing to do' (Williams, 2004). 'Family' in the context of care is not (only) about *who* but about *how* care is thought about. For the provision of ideal family care, one's own home has important significance as the nexus of intimate relationships. In Chapter 3, I discussed the geographies of care, with a focus on the utopia of home and its opposite, the institutional setting. I have argued that the dichotomy between loving, affectionate caring and professionalised, institutionalised work is reproduced in the construction of the physical place. I have shown that people continuously express a preference for being cared for at home and that institutional care arrangements are thought of as quintessential places, which lack intimacy and thus care. Similar to the notion of family, home subsumes certain values, virtues and aspects of social life, which are constructed in opposition to the dominant, hegemonic market logic. In that sense, home represents an image that is both nostalgic (as it entails traditional family ideals) and progressive (in opposition to a marketised world). People's desire for home, as both a symbol and a physical space, goes far beyond an uncritical favouring of traditional family structures and ways of living; nostalgia also represents the rejection of an exclusively economic logic. Aspects of nostalgia were further explored in Chapter 4, where I described the construction of community as an ideological extension of family and the neighbourhood as based on a nostalgic imagination of ideal caring situations. Combined with the safe space of home and the framework of the family, community is constructed as a counterforce to what is perceived as hostile, individualising and pressuring economic, political and social developments. Community, I argued, represents ideals, emotions and desires about care in the broader society. Similar to the traditional family, community is felt to be under threat from economic developments and it is located as having taken place at other times or in other places.

Blurring and bridging the dichotomies

Throughout this book, I have tried to disentangle the discursive patterns that underlie the moral foundation of the meaning of care. Each component of people's associations with care can be analysed as being constituted by dichotomies that distinguish positive from negative associations, (morally) good from bad practices and idealisations from dystopias. While the binary opposites frame the main meaning, it can be shown that, empirically, the boundaries constructing the dichotomies are continuously and recurrently blurred, bridged and challenged. These challenges to the mainstream discursive constructions, I have argued, need to be explored and recognised, since the organisation and experience of empirical caring practices need to be understood as moving between and beyond dichotomies. Additionally, bridging and challenging the boundaries is also normatively and politically important for the organisation of fair and ethical care. If 'ideal' care is imagined as a state of loving, devotional minding, for example, it is important to be conscious of the potential reductions to gender stereotypes in this context. Migrant women in particular are constructed as 'the other', and they represent the ideal of a caring identity. In that sense, real caring, provided by migrant women in people's houses, is praised as representing the general ideal of being a morally good person. In the introduction, I further stated that those involved in the provision of care are facing feminisation due to the moral and ideological construction of care. The ethics of a care approach helped to identify that many of the associations with the ideal carer identity reflect a stereotypical feminised identity. A distinction between the caring, feminised subject and the career-driven masculine subject is established. Interestingly, men and women can fulfil both roles in this discourse, for example when Austrian women are described as being *business driven*, in comparison to Slovakian women who still *care*. The gender connotations remain, however, as the dependency relations are seen to create clear boundaries between male and female characteristics.

The dichotomies of care and the boundaries between those who care and those who do not are also present in the discussion of dependency relations themselves. In Chapter 5, I showed that old age is frequently associated with dependency, passivity and suffering. I argued that people express anxieties about dependency and vulnerability and that a dichotomy of the independent, self-sufficient actor on the one hand and the dependent, vulnerable, elderly care receiver on the other is created. I argued that desiring and imagining the ideal of independent living as long as possible sketches an ideology that contradicts many of the values of care, and any divergence from the independent actor's role is seen as being inferior to the ideal. Dependency is, furthermore, linked to life in institutional settings, in which people are left to others' goodwill, while the home, on the other hand, is constructed in the sense that it allows independent living. Chapter 6 took up the theme of the discursive construction and totalisation of clear identities of dependent or independent actors and merged it with another theme running through the whole book, the limits and challenges of the professionalisation and commodification of care. I argued that care is ideologically and morally positioned in opposition to work, employment, politics, bureaucracy and markets. The aversions to institutionalisation, professionalisation, instrumentality and politics lead to the construction of a dichotomy of care and work through which care is positioned in opposition to the world of markets, paid work and economic individualism. The 'price-lessness' of care is seen as a gift, both to the elderly and to society. Praising unpaid carers as morally superior places them outside normal citizenship and affects them in any choices they make, simply because the discourse around care presents them as morally superior. The commodification or professionalisation of care is difficult, since the logic of the market challenges the moral meaning of care. Two 'hostile worlds' are created in the discursive construction of care, and this dichotomy is reproduced in narratives, ideals and opinions. Additionally, I argued for a rethinking of carers' identities. The sphere of markets is identified with the buying and selling of services and specific tasks, while the sphere of care refers to the presence and commitment of people close to one. Importantly, this book has also shown that there is some progressive value in care discourses and in the discursive use (and challenges) of dichotomies. On the one hand, the construction of care reproduces the marginalised and vulnerable position of both carers and cared for by focusing on the idea of 'being there for each other'. On the other hand, however, this discursive practice also constructs care as a counter-discourse. Community, for example, also needs to be seen as a counter-discourse within an individualised, economised and marketised world, and I have also pointed towards the positive aspects of non-traditional forms of care and responsibility for care. A tendency can be identified to see community as a modern answer to the demands of care for elderly people; community might in that sense replace traditional forms of family responsibilities. Nostalgia also includes an acknowledgement that traditional forms of living cannot be brought back; hence, community describes something of a contrast to the marketisation and individualisation of modern life. Similarly, the construction and the fear of dependency and the challenges of commodified and professionalised care demonstrate a partial acceptance and acknowledgement of relating and being there for each other and show some criticism of an ideal of individualisation and self-dependency.

Political implications

I also asked whether the moral conception of care can be contested and challenged, and throughout the book I have identified potential counter-discourses, ambivalences and tensions in people's accounts. At the same time, the question arose of what political potential the analysis of this discourse might have. Norms and values in society can be influenced and changed by political action in an attempt to combine a focus on the moral construction of meaning with the socio-economic context. The idea that collective action can and should lead to changes in the moral framework of society (see Rosenbeck, 1998) points to the possibilities for the active creation of the moral meaning of care. One example of how political intervention can change the moral grammar of a society is the Scandinavian Social Democratic discussion of gender (see Karlsson, 1998; Siim, 1987, 1993; Wærness, 1998) whereby governments implemented various policies with respect to gender equality, care responsibilities and a distinction between the public and the private in everyday life (see Rosenbeck, 1998). It is argued (Sörensen and Bergqvist, 2002) that the policies on issues of care and gender equality were also meant to promote the abandonment of traditional roles and identities of men and women. This implies an explicit idea of social change through political intervention in which the notion of ' "gender" is conceptually transformed from a synonym for "divisions of work" into a synonym for "values" and "interests" ' (Skjeie and Siim, 2000: 354). As Hernes (1987; also see Skjeie and Siim, 2000 on the importance of social movements and Sörensen and Bergqvist, 2002 on women's mobilisation) convincingly shows, the intended change was explicitly based on a notion of the subjects acting in two ways: women's agitation ('feminization from below') and a change in government policy ('state feminism from above'). For the discussion of the meaning of care, this means that a moral grammar can be challenged by political intervention. At the same time, however, other spheres of public discourses need to be considered in order to enable a reconfiguration of people's understanding of care.

Another aim of this book was to explore the paradox that care is valued very highly but marginalised politically. In Chapter 6, I discussed people's ambivalent position vis-à-vis the role of politics. On the one hand, politics is seen as interfering with the 'natural' provision and organisation of care; on the other hand, political intervention is desired to enable and secure the very natural construction and organisation of care. I would argue that, again, a rethinking of the dichotomy of care and work needs to be attempted. If economic exchange and intimate, loving involvement are not understood as contradictions anymore, political interventions can create new arrangements for those caring and those being cared for. It is inevitable not only for those involved in care but also for society, in general, to put care and intimacy onto the political agenda. This book has tried to show that care and politics, care and work and intimacy and markets are seen and constructed as moral and ideological opposites. Care represents an ideal world, but it is a world that contradicts the dominant, hegemonic focus. This is a broad reproduction of the public–private distinction, of the ethics of care against a work ethic, of feminisation against the dominant masculine ideology; it also offers, however, the potential to intervene in the discourses, which shape dominant societal arrangements. People's desires to be there for each other can be seen as a starting point for a political ethics of care. The consciousness needs to be raised that taking care out of the private realm and making it a central focus for the public world need not mean to reconstruct it under the umbrella of marketisation and commodification. A public focus on relating and being there for each other needs to be reconstructed to allow attention to be paid to the negative and potentially exploitative consequences of traditional care arrangements while, at the same time, acknowledging and valuing the desires and feelings people associate with 'ideal care'.

Care is constructed as intimacy, as a feeling, as love and an ideal of 'being there for each other'. In Chapter 6, I argued that the construction of these two hostile worlds needs to be challenged, reconciled and overcome. A combination of financial transactions with intimate care is happening, and, theoretically, asymmetrical relationships need not be harmfully hierarchical. In that sense, professionalism and employment do not necessarily mean non-attachment, or a non-intimate relationship. The political recognition of the construction of dichotomies and problems associated with this construction is needed. Care should not be seen as an opposite to work and markets but as a prerequisite for their existence. Currently, in many ways, care is only recognised when it can be described as 'work'. Carers are recognised for their contribution, as they perform useful work for society. I would propose, however, that the exclusive focus on work itself is the problem and care needs to be recognised as going beyond it, as a practice reflecting the desires and wishes in and of society. Not only because it involves work but also because it is a deeply human practice, care needs to be recognised and valued. Iris Marion Young (2002: 55) lucidly captures this fundamental rethinking of social contributions in her description of meaningful work:

> An ideal of meaningful work says that [the] work people do ought to be clearly connected to social uses and should be recognized by others for its contribution to the well-being of persons or their dwelling environments or to the well-being of other creatures and their dwelling environments.

The construction of care in contrast to the dominant market logic can foster a progressive approach to a new understanding of organising social practices and relations. What is needed is a new language and a new moral discourse, which enable a combination of loving and dedicated 'being there for each other' with the values of equality, autonomy and justice. A new ethics of care, seeking to bridge the demands of a focus on relationality and individual justice, would help, on the one hand, those involved with and confronted by caring relations and, on the other, society in general. The paradoxical situation that people idealise care as being there for each other, imagined as a process based on love and intimacy, while at the same time being anxious about becoming dependent on others, can similarly only be resolved by understanding what people associate with care and by valuing what care really means to them and to society. If the ideal of care (and its associated moral expression) becomes a dominant understanding in society, then the tensions and contradictions in people's feelings and emotions can be addressed. For that, the discussion in Chapter 5 is particularly important. Independence in the sense of self-sufficiency as an ideological ideal needs to be challenged and replaced by an acknowledgement of mutual, and sometimes not mutual, dependencies on each other. Young (2002) argues that self-sufficiency is impossible but seductive. She proposes a different understanding of autonomy in which 'forms of dependence and interdependence [...] should be understood as *normal* conditions of being autonomous' (Young, 2002: 47). Autonomy, understood in this way, requires social support and recognition of our dependency on each other. A political and discursive intervention starting with the recognition of dependency would enable an approach which fulfils people's desire for someone providing loving, intimate 'being there for each other', but at the same time does creating vulnerability to exploitation.

Sociopolitical interventions in care need to be aware of the discursive constructions that constitute the meaning of certain practices around care for people. This book wants to contribute to a rethinking of the dichotomies that strongly frame and shape all political, economic and social actions in relation to the organisation of care for elderly people. Writing this book has had a considerable impact on my own understanding of care and my own experiences of the questions and practices of age, illness, dependency and care. Writing a book on the meaning of care should not remain an exclusively academic experience; rather, I want to contribute explicitly to the questioning and reframing of the discursive constructions that underlie the current constitution of care. In a very moving contribution to the relationship between the personal and the professional, Eva Kittay (2009) reflects on the links between her own academic position as a philosopher and her role as the mother of a disabled child:

> My daughter, Sesha, will never walk the halls of academe, but when what happens within these halls has the potential to affect her, then I as an academic have an obligation to socialize academe to accept my daughter. Such 'care' may seem to be far from the daily care that her fully dependent body requires, and it may appear to be far-fetched to call this 'care', but it is part and parcel of that labor of love that we do as parents [...] more still in the case of those who are so disabled that they cannot speak for themselves.
>
> (Kittay, 2009: 611)

Care is a deeply relational experience and the meaning of care constitutes itself based on the way we see both personal and intimate, and broader and societal, relations. By participating in concrete caring practices, but also in private, public and academic discourses, we need to be aware of our role in producing and reproducing the meaning of care and the consequences for those requiring, and those providing, care.

Notes

1 Introduction

1. This position can be compared to Christian Ethics in which a sin or good action is defined by its motivation and moral character (see Žižek, 2000).
2. This is a brief explanation of the essence of Darwall's (2002) 'Rational Care Theory of Welfare'.
3. For a discussion of postmodern ethics in the context of (health)care, see Fox (2000).
4. Here, an interesting similarity can be found with other religiously inspired writings. Buber (1958), for example, emphasises that a relationship of love between I and Thou is characterised by the responsibility of I for Thou. Consequently, I is only constituted in relation to Thou, and eventually to the external Thou, God (Buber, 1958: 75).
5. Bahr and Bahr also raise this point with respect to self-sacrifice in a Levinasian sense when they hold that '[s]elf-sacrifice tends not to be the result of conscious, rational decision making. Its voluntary nature is more reflexive than cognitive, more a matter of community identity, intuition, and reaction, than a realistic weighing of alternatives. It is a response to need, not an assessment of possible damage to one's personal projects' (Bahr and Bahr, 2001: 1250–1251).
6. Held (2002b) acknowledges that a focus on bodies, emotions, embodiment and so on (i.e. naturalism) can be attractive to feminist theory and that it can also, however, be a dangerous path to essentialisation.
7. But Held (2002b) also immediately states that this ethics of care must be distant from a non-naturalistic Kantian morality, as Kantian theories are unsuitable to deal with experiences of family and friendship.
8. My interpretation of materialism is not based on a Darwinian ethics (see Blackledge and Kirkpatrick, 2002), nor does it refer to the narrow conception Williams (1980) sketches out with its emphasis on prioritising nature over the mind; it rather resembles a tradition of Critical Theory influenced by the early Frankfurt School whose basic thesis is that human being is dependent on the overall constitution of the world (Horkheimer, 1980) in its material, cultural and historical setting.
9. Honneth (1995) does not interpret 'consensus' in a homogenous sense. It should rather be understood as the outcome of collectively constructed moral expectations based on individuals' struggles.
10. In this context, Fraser (2003a) also argues against a simple poststructuralist anti-dualism.
11. Here, Fraser questions Habermas' (1989) relative static conception and terminology of the spheres of system and lifeworld.
12. Clearly, some of these terms can have negative connotations for some people. The notion of family, for example, can represent negative experiences for some people or an old-fashioned institution for societal organisation

for others. However, even those people have to be conscious of the mainstream connotations of family as a positive concept. The terminology of dependency and independence is another important example in this context.

2 Who Should Care? The Construction of Caring Relationships

1. Interestingly, Barker also notes that, in particular, younger male carers (under 45) used kin terms. She argues that '[t]his is just one strategy by which they normalize their otherwise suspect relationship with their mainly very elderly female dependents' (2002: 165).

5 Who Is Seen to Be Cared for? The Construction of the Care Receiver

1. My own translation.

6 Buying and Selling Care? The Intrusion of Markets and Bureaucracy

1. In Chapter 5, I made a reference to the concept of 'social death', which Tim refers to at the end of this extract.
2. Besides 'love labour', the other forms of care that Lynch (2007) describes in her work are 'general care labour' and 'solidarity work'.
3. Nelson and England (2002:1), besides others, argue for a move away from a dualistic view that women, love, altruism and the family are radically different to man, rationality, the market and work.

Bibliography

Aboim, S (2010), Gender cultures and the division of labour in contemporary Europe: a cross-national perspective, *Sociological Review*, Vol. 58, pp. 171–197.

Abrahamson, P (1999), The welfare modelling business, *Social Policy & Administration*, Vol. 33, (4), pp. 394–415.

Ahmed, S, Castañeda, C, Fortier, A and Sheller, M (2003), Introduction: Uprooting/Regroundings: Questions of Home and Migration, In Ahmed, S, Castañeda, C, Fortier, A and Sheller, M (eds.), *Uprootings/Regroundings: Questions of Home and Migration*, Oxford: Berg, pp. 1–19.

Aldridge, M (1994), *Making Social Work News*, London: Routledge.

Alexander, J C (1993), Citizen and Enemy as Symbolic Classification: On the Polarizing Discourse of Civil Society, In Fournier, M and Lamont, M (eds.), *Where Culture Talks: Exclusion and the Making of Society*, Chicago: Chicago University Press, pp. 289–308.

Alexander, J C (2006), *The Civil Sphere*, Oxford: Oxford University Press.

Alleyne, B (2002), An idea of community and its discontents: towards a more reflexive sense of belonging in multicultural Britain, *Ethnic and Racial Studies*, Vol. 25, (4), pp. 607–627.

Améry, J (1968), *Über das Altern: Revolte und Resignation*, Stuttgart: Ernst Klett Verlag.

Anderson, B (2000), *Doing the Dirty Work? The Global Politics of Domestic Labour*, London: Zed Books.

Anderson, B R (1991), *Imagined Communities: Reflections on the Origin and Spread of Nationalism*, revised edition, London: Verso.

Andrews, M (1999), The seductiveness of agelessness, *Ageing & Society*, Vol. 19, pp. 301–318.

Anttonen, A and Zechner, M (2011), Theorising Care and Care Work, In Pfau-Effinger, B and Rostgaard, T (eds.), *Care, Work and Welfare in Europe*, Houndmills: Palgrave Macmillan, pp. 15–34.

Araujo Sartorio, N de and Pavone Zoboli, E (2010), Images of a 'good nurse' presented by teaching staff, *Nursing Ethics*, Vol. 17, (6), pp. 687–694.

Augé, M (1995), *Non-Places: Introduction to an Anthropology of Supermodernity*, London: Verso.

Bach, S, Kessler, I and Heron, P (2012), Nursing a grievance? The role of healthcare assistants in a modernized national health service, *Gender, Work and Organization*, Vol. 19, (2), pp. 205–224.

Bachelard, G (1969), *The Poetics of Space*, Boston: Beacon Press.

Badelt, C and Österle, A (2001), *Grundzüge der Sozialpolitik: Sozialpolitik in Österreich: Spezieller Teil*, 2nd ed., Vienna: Manz Verlag.

Badelt, C, Holzmann-Jenkins, A, Matul, C and Österle, A (1997), *Analyse der Auswirkungen des Pflegevorsorgesystems*, Research report on behalf of the Bundesministerium für Arbeit, Gesundheit und Soziales, Vienna.

Bahr, H M and Bahr, K S (2001), Families and self-sacrifice: Alternative models and meanings for family theory, *Social Forces*, Vol. 79, (4), pp. 1231–1258.

Bakan, A and Stasiulis, D (1997), Introduction, In Bakan, A and Stasiulis, D (eds.), *Not One of the Family: Foreign Domestic Workers in Canada*, Toronto: University of Toronto Press, pp. 3–28.

Baldwin, M (2000), *Care Management and Community Care: Social Work Discretion and the Construction of Policy*, Aldershot: Ashgate.

Baldwin, S (1995), Love and Money: The Financial Consequences of Caring for an Older Relative, In Allen, I and Perkins, E (eds.), *The Future of Family Care for Older People*, London: HMSO, pp. 119–140.

Baltes, P and Smith, J (2003), New frontiers in the future of aging: From successful aging of the young old to the dilemmas of the fourth age, *Gerontology*, Vol. 49, pp. 123–135.

Barker, J C (2002), Neighbors, friends, and other nonkin caregivers of community-living dependent elders, *Journal of Gerontology*, Vol. 57B, (3), pp. 158–167.

Barnes, C, Marcer, G and Shakespeare, T (1999), *Exploring Disability: A Sociological Introduction*, Cambridge: Polity Press.

Barrera, A (2008), Globalization's shifting economic and moral terrain: contesting marketplace mores, *Theological Studies*, Vol. 69, (2), pp. 290–308.

Bartmanski, D (2011), Successful icons of failed time: Rethinking post-communist nostalgia, *Acta Sociologica*, Vol. 54, (3), pp. 213–231.

Bauld, L, Chesterman, J, Davies, B, Judge, K and Mangalore, R (eds.) (2000), *Caring for Older People: An assessment of Community Care in the 1990s*, Aldershot: Ashgate.

Bauman, Z (1993), *Postmodern Ethics*, Oxford: Blackwell Publishers.

Bauman, Z (1995), *Life in Fragments: Essays in Postmodern Morality*, Oxford: Blackwell.

Bauman, Z (2001), *Community: Seeking Safety in an Insecure World*, Cambridge: Polity Press.

Bauman, Z (2004), *Identity: Conversations with Benedetto Vecchi*, Cambridge: Polity Press.

Beauvoir, S de (1972), *The Coming of Age*, translated by P. O'Brian, New York: Putnam's Sons.

Beck, U (1998), *Democracy Without Enemies*, Cambridge: Polity Press.

Beck, U (2009), *World at Risk*, Cambridge: Polity Press.

Beck, U and Beck-Gernsheim, E (2001), *Individualization: Institutionalised Individualism and Its Social and Political Consequences*, London: Sage Publications.

Bednarek, M (2006), *Evaluation in Media Discourse: Analysis of a Newspaper Corpus*, London: Continuum.

Bengtson, V and Roberts, R (1991), Intergenerational solidarity in aging families: An example of formal theory construction, *Journal of Marriage and the Family*, Vol. 53, (4), pp. 856–870.

Bengtson, V, Giarrusso, R, Mabry, J B and Silverstein, M (2002), Solidarity, conflict, and ambivalence: Complementary or competing perspectives on intergenerational relationships? *Journal of Marriage and the Family*, Vol. 64, (3), pp. 568–576.

Bertram, H (2002), Intimität, Ehe, Familie und private Beziehungen, *Soziale Welt: Zeitschrift für sozialwissenschaftliche Forschung und Praxis*, Vol. 53, (4), 415–422.

Bettio, F, Simonazzi, A and Villa, P (2006), Change in care regimes and female migration: the 'care drain' in the Mediterranean, *Journal of European Social Policy*, Vol. 16, (3), pp. 271–285.

Biggs, S (2001), Toward critical narrativity: Stories of aging in contemporary social policy, *Journal of Aging Studies*, Vol. 15, (4), pp. 303–316.

Billig, M (2003), Critical Discourse Analysis and the Rhetoric of Critique, In Weiss, G and Wodak, R (eds.), *Critical Discourse Analysis: Theory and Interdisciplinarity*, Basingstoke: Palgrave Macmillan, pp. 35–46.

Binstock, R (2005), Old-age policies, politics, and ageism. *Generations*, Vol. 29, (3), pp. 73–78.

Blackledge, P and Kirkpatrick, G (eds.) (2002), *Historical Materialism and Social Evolution*, Basingstoke: Palgrave Macmillan.

Bloor, M, Frankland, J, Thomas, M and Robson, K (2001), *Focus Groups in Social Research*, London: Sage Publications.

Blunt, A (2003), Collective memory and productive nostalgia: Anglo-Indian homemaking at McCluskieganj, *Environment and Planning D: Society and Space*, Vol. 21, pp. 717–738.

Blunt, A (2005), Cultural geography: Cultural geographies of home, *Progress in Human Geography*, Vol. 29, (4), pp. 505–515.

Blunt, A and Dowling, R (2006), *Home*, Abingdon: Routledge.

Bobbio, N (2001), Old Age, In Bobbio, N (ed.), *Old Age and Other Essays*, Cambridge: Polity Press, pp. 3–31.

Bondi, L (2008), On the relational dynamics of caring: a psychotherapeutic approach to emotional and power dimensions of women's care work, *Gender, Place & Culture*, Vol. 15, (3), pp. 249–265.

Bonnett, A (2010), *Left in the Past: Radicalism and the Politics of Nostalgia*, New York: Continuum.

Bookman, A (2008), Innovative models of aging in place: Transforming our communities for an aging population, *Community, Work & Family*, Vol. 11, (4), pp. 419–438.

Bornat, J, Johnson, J, Pereira, C, Pilgrim, D and Williams, F (eds.) (1997), *Community Care: A Reader*, London: Macmillan.

Bowlby, S, McKie, L, Gregory, S and MacPherson, I (2010), *Interdependency and Care over the Lifecourse*, London: Routledge.

Boym, S (2001), *The Future of Nostalgia*, New York: Basic Books.

Brand, K (2009), *The Care Factory*, Film Documentary, 16 minutes, Netherlands.

Brenner, J and Haaken, J (2000), Utopian thought: Re-visioning gender, family, and community, *Community, Work & Family*, Vol. 3, (3), pp. 333–347.

Brown, M (2003), Hospice and the spatial paradoxes of terminal care, *Environment and Planning A*, Vol. 35, pp. 833–851.

Bubeck, D E (1995), *Care, Gender, and Justice*, Oxford: Oxford University Press.

Bubeck, D E (2002), Justice and the Labor of Care, In Kittay, E F and Feder, E K (eds.), *The Subject of Care: Feminist Perspectives on Dependency*, Lanham: Rowner & Littlefield, pp. 160–185.

Buber, M (1958), *I and Thou*, New York: Collier Books.

Burkart, G (2002), Stufen der Privatheit und die discursive Ordnung der Familie, *Soziale Welt: Zeitschrift für sozialwissenschaftliche Forschung und Praxis*, Vol. 53, (4), pp. 397–414.

Cangiano, A, Shutes, I, Spencer, S and Leeson, G (2009), *Migrant Care Workers in Ageing Societies: Research Findings in the United Kingdom*, Executive Summary, Oxford: COMPAS.

Castells, M (1997), *The Power of Identity, The Information Age: Economy, Society and Culture*, Vol. 2, Oxford: Blackwell Publishers.

Caygill, H (2002), *Levinas & the Political*, London: Routledge.

Ceci, C, Björnsdóttir, K and Purkis M (2012), Introduction: Home, Care, Practice – Changing Perspectives on Care at Home for Older People, In Ceci, C, Björnsdóttir, K and Purkis M (eds.), *Perspectives on Care at Home for Older People*, New York: Routledge, pp. 1–22.

Chantler, K (2006), Independence, dependency and interdependence: Struggles and resistances of minoritized women within and on leaving violent relationships, *Feminist Review*, Vol. 82, pp. 27–49.

Chouliaraki, L and Fairclough, N (1999), *Discourse in Late Modernity: Rethinking Critical Discourse Analysis*, Edinburgh: Edinburgh University Press.

Christie, A (2006), Negotiating the uncomfortable intersections between gender and professional identities in social work, *Critical Social Policy*, Vol. 26, (2), pp. 390–411.

Clarke, L (1995), Family Care and Changing Family Structures: Bad News for the Elderly? In Allen, I and Perkins, E (eds.), *The Future of Family Care for Older People*, London: HMSO, pp. 19–49.

Conradson, D (2003a), Geographies of care: spaces, practices, experiences, *Social & Cultural Geography*, Vol. 4, (4), pp. 451–454.

Condradson, D (2003b), Spaces of care in the city: The place of a community crop-in centre, *Social & Cultural Geography*, Vol. 4, (4), pp. 507–525.

Coontz, S (1992), *The Way We Never Were: American Families and the Nostalgia Trap*, New York: Basic Books.

Coupland, J (2009), Discourse, identity and change in mid-to-late life: interdisciplinary perspectives on language and ageing, *Ageing & Society*, Vol. 29, pp. 849–861.

Cowen, H (1999), *Community Care, Ideology and Social Policy*, Harlow: Prentice Hall Europe.

Daatland, S (1990), 'What are Families For?': On family solidarity and preference for help, *Ageing & Society*, Vol. 10, (1), pp. 1–15.

Daatland, S (1997), Family solidarity, popular opinion and the elderly: Perspectives from Norway, *Ageing International*, Vol. 24, (1), pp. 51–62.

Da Roit, B (2010), *Care Strategies: Changing Elderly Care in Italy and the Netherlands*, Amsterdam: Amsterdam University Press.

Da Roit, B and Le Bihan, B (2010), Similar and yet so different: Cash-for-care in six European countries' long-term care policies, *Milbank Quarterly*, Vol. 88, (3), pp. 286–309.

Da Roit, B and Weicht, B (2013), Migrant care work and care, migration and employment regimes: A fuzzy-set analysis, *Journal of European Social Policy*, Vol. 23, (5), pp. 469–486.

Da Roit, B, Hoogenboom, M and Weicht, B, The Gender Informal Care Gap. A Fuzzy Set Analysis of Cross-Country Variations, unpublished paper.

Dalley, G (1996), *Ideologies of Caring: Rethinking Community and Collectivism*, 2nd ed., Basingstoke: MacMillan Press.

Daly, M (2002), Care as a good for social policy, *Journal of Social Policy*, Vol. 31, (2), pp. 251–270.

Daly, M and Lewis, J (2000), The concept of social care and the analysis of contemporary welfare states, *British Journal of Sociology*, Vol. 51, (2), pp. 281–298.

Darwall, S (2002), *Welfare and Rational Care*, Woodstock: Princeton University Press.

Davis, F (1979), *Yearning for Yesterday: A Sociology of Nostalgia*, New York: The Free Press.

De Certeau, M, Giard, L and Mayol, P (1998), *The Practice of Everyday Life, Vol. 2: Living and Cooking*, Minneapolis: University of Minnesota Press.

de Cillia, R, Reisigl, M and Wodak, R (1999), The discursive construction of national identities, *Discourse & Society*, Vol. 10, (2), pp. 149–173.

Dean, H (2004), Reconceptualising Dependency, Responsibility and Rights, In Dean, H (ed.), *The Ethics of Welfare: Human Rights, Dependency and Responsibility*, Bristol: The Policy Press, pp. 193–209.

Dean, H and Rogers, R (2004), Popular Discourses of Dependency, Responsibility and Rights, In Dean, H (ed.), *The Ethics of Welfare: Human Rights, Dependency and Responsibility*, Bristol: The Policy Press, pp. 69–88.

Degiuli, F (2007), A job with no boundaries: Home eldercare work in Italy, *European Journal of Women's Studies*, Vol. 14, (3), pp. 193–207.

Dench, G, Gavron, K and Young, M (2006), *The New East End: Kinship, Race and Conflict*, London: Profile Books.

Derrida, J (1981), *Positions*, Chicago: University of Chicago Press.

Doheny, S (2004), Responsibility and Welfare: In Search of Moral Sensibility, In Dean, H (ed.), *The Ethics of Welfare: Human Rights, Dependency and Responsibility*, Bristol: The Policy Press, pp. 49–66.

Drakeford, M (2006), Ownership, regulation and the public interest: The case of residential care for older people, *Critical Social Policy*, Vol. 26, (4), pp. 932–944.

Duyvendak, J (2011), *The Politics of Home: Belonging and Nostalgia in Western Europe and the United States*, Basingstoke: Palgrave Macmillan.

Easthope, H (2004), A place called home, *Housing, Theory and Society*, Vol. 21, (3), pp. 128–138.

Egdell, V, Bond, J, Brittain, K and Jarvis, H (2010), Disparate routes through support: Negotiating the sites, stages and support of informal dementia care, *Health & Place*, Vol. 16, pp. 101–107.

Ellis, K (2004), Dependency, Justice and the Ethic of Care, In Dean, H (ed.), *The Ethics of Welfare: Human Rights, Dependency and Responsibility*, Bristol: The Policy Press, pp. 29–48.

Ellis, K and Rogers, R (2004), Fostering a Human Rights Discourse in the Provision of Social Care for Adults, In Dean, H (ed.), *The Ethics of Welfare: Human Rights, Dependency and Responsibility*, Bristol: The Policy Press, pp. 89–109.

Evetts, J (2003), The sociological analysis of professionalism: Occupational change in the modern world, *International Sociology*, Vol. 18, (2), pp. 395–415.

Evetts, J (2011), Sociological analysis of professionalism: Past, present and future, *Comparative Sociology*, Vol. 10, pp. 1–37.

Fealy, G, McNamara, M, Treacy, M and Lyons, I (2011), Constructing ageing and age identities: A case study of newspaper discourses, *Ageing & Society*, Vol. 31, pp. 1–18.

Feder, E K and Kittay, E F (2002), Introduction, In Kittay, E F and Feder, E K (eds.), *The Subject of Care: Feminist Perspectives on Dependency*, Lanham: Rowner & Littlefield, pp. 1–13.

Finch, J (1995), Responsibilities, Obligations and Commitments, In Allen, I and Perkins, E (eds.), *The Future of Family Care for Older People*, London: HMSO, pp. 51–64.

Finch, J and Mason, J (1990), Filial obligations and kin support for elderly people, *Ageing and Society*, Vol. 10, pp.151–176.

Finch, J and Mason, J (2000), Filial Obligations and Kin Support for Elderly People, In Gubrium, J F and Hostein, J A (eds.), *Ageing and Everyday Life*, Oxford: Blackwell Publishing, pp. 193–213.

Fine, M (2005), Individualization, risk and the body: Sociology and care, *Journal of Sociology*, Vol. 41, (3), pp. 247–266.

Fine, M and Glendinning, C (2005), Dependence, independence or interdependence? Revisiting the concepts of 'care' and 'dependency', *Ageing & Society*, Vol. 25, pp. 601–621.

Fineman, M L A (2002), Masking Dependency: The Political Role of Family Rhetoric, In Kittay, E F and Feder E K (eds.), *The Subject of Care: Feminist Perspectives on Dependency*, Lanham: Rowman & Littlefield, pp. 215–244.

Fink, J (2004), Questions of Care, In Fink, J (ed.), *Care: Personal Lives and Social Policy*, Bristol: The Policy Press, pp. 1–42.

Firth, A (2007), Transcending the 'merely material': Secular morality and progressive politics, *History of the Human Sciences*, Vol. 20, (1), pp. 67–81.

Fitzpatrick, T (2008), *Applied Ethics and Social Problems: Moral Questions of Birth, Society and Death*, Bristol: The Policy Press.

Fortier, A (2003), Making Home: Queer Migrations and Motions of Attachment, In Ahmed, S, Castañeda, C, Fortier, A and Sheller, M (eds.), *Uprootings/Regroundings: Questions of Home and Migration*, Oxford: Berg, pp. 115–136.

Foucault, M (1972), *The Archaeology of Knowledge*, London: Routledge.

Fox, N (2000), The Ethics and Politics of Caring: Postmodern Reflections, In Williams, S J, Gabe, J and Calnan, M (eds.), *Health, Medicine and Society: Key Theories, Future Agendas*, London: Routledge, pp. 333–349.

Fraser, N (1989), *Unruly Practices: Power, Discourse and Gender in Contemporary Social Theory*, Cambridge: Polity Press.

Fraser, N (1990), Rethinking the Public Sphere: A Contribution to the Critique of Actually Existing Democracy, *Social Text*, No. 25/26, pp. 56–80.

Fraser, N (2000), Rethinking Recognition, *New Left Review*, 3, pp. 107–120.

Fraser, N (2003a), Social Justice in the Age of Identity Politics: Redistribution, Recognition, and Participation, In Fraser, N and Honneth, A, *Redistribution or Recognition? A Political-Philosophical Exchange*, London: Verso, pp. 7–109.

Fraser, N (2003b), Distorted Beyond All Recognition: A Rejoinder to Axel Honneth, In Fraser, N and Honneth, A (eds.), *Redistribution or Recognition? A Political-Philosophical Exchange*, London: Verso, pp. 198–236.

Fraser, N (2005), Reframing Justice in a Globalizing World, *New Left Review*, 36, pp. 69–88.

Fraser, N (2008), *Scales of Justice: Reimagining Political Space in a Globalizing World*, Cambridge: Polity Press.

Fraser, N and Gordon, L (1994), A genealogy of *dependency*: Tracing a keyword of the U.S. Welfare State, *Signs*, Vol. 19, (2), pp. 309–336.

Friedan, B (1993), *The Fountain of Age*, London: Vintage.

Froggatt, K (2001), Life and death in English nursing homes: Sequestration or transition? *Ageing and Society*, Vol. 21, (3), pp. 319–332.

Gal, S (2004), A Semiotics of the Public/Private Distinction, In Scott, J W and Keates, D (eds.), *Going Public: Feminism and the Shifting Boundaries of the Private Sphere*, Urbana: Univ. of Illinois Press, pp. 261–277.

Garland-Thompson, R (2005), Feminist Disability Studies, *Signs*, Vol. 30, (2), pp. 1557–1587.

Giddens, A (1991), *Modernity and Self-Identity: Self and Society in the Late Modern Age*, Cambridge: Polity Press.

Giddens, A (1998), *The Third Way: The Renewal of Social Democracy*, Cambridge: Polity Press.

Gilleard, C and Higgs, P (2011), Ageing abjection and embodiment in the fourth age, *Journal of Aging Studies*, Vol. 25, pp. 135–142.

Gilligan, C (1982), *In a Different Voice: Psychological Theory and Women's Development*, London: Harvard University Press.

Gilligan, C (1993), Reply to Critics, In Larrabee, M J (ed.), *An Ethic of Care: Feminist and Interdisciplinary Perspectives*, London: Routledge, pp. 207–214.

Glendinning, C (2008), Increasing choice and control for older and disabled people: A critical review of new developments in England, *Social Policy & Administration*, Vol. 42, (5), pp. 451–469.

Glendinning, C, Challis, D, Fernández, J, Jacobs, S, Jones, K, Knapp, M, Manthorpe, J, Moran, N, Netten, A, Stevens, M and Wilberforce, M (2008), *Evaluation of the Individual Budgets Pilot Programme: Summary Report*, York: Social Policy Research Unit.

Glucksmann, M (2005), Shifting Boundaries and Interconnections: Extending the 'Total Social Organisation of Labour', In Pettinger, L, Parry, J, Taylor, R and Glucksmann, M (eds.), *A New Sociology of Work?* Oxford: Blackwell Publishing, pp. 19–36.

Glucksmann, M and Lyon, D (2006), Configurations of care work: Paid and unpaid elder care in Italy and the Netherlands, *Sociological Research Online*, Vol. 11, (2), http://www.socresonline.org.uk/11/2/glucksmann.html.

Godden, J and Helmstadter, C (2004), Woman's mission and professional knowledge: Nightingale nursing in colonial Australia and Canada, *Social History of Medicine*, Vol. 17, (2), pp. 157–174.

Greener, I (2004), The three moments of New Labour's health policy discourse, *Policy & Politics*, Vol. 32, (3), pp. 303–316.

Grenier, A and Hanley, J (2007), Older women and 'frailty': Aged, gendered and embodied resistance, *Current Sociology*, Vol. 55, (2), pp. 211–228.

Groenhout, R (1998), Care theory and the ideal of neutrality in public moral discourse, *Journal of Medicine and Philosophy*, Vol. 23, (2), pp. 170–189.

Groenhout, R (2003), *Theological Echoes in an Ethic of Care*, Occasional Papers of the Erasmus Institute, 2.

Groenhout, R (2004), *Connected Lives: Human Nature and an Ethics of Care*, Oxford: Rowman & Littlefield.

Guberman, N, Maheu, P and Maillé, C (1992), Women as family caregivers: Why do they care? *The Gerontologist*, Vol. 32, (5), pp. 607–617.

Haberkern, K and Szydlik, M (2010), State care provision, societal opinion and children's care for older parents in 11 European countries, *Ageing & Society*, Vol. 30, pp. 299–323.

Habermas, J (1989), *The Structural Transformation of the Public Sphere: An Inquiry into a Category of Bourgeois Society*, Cambridge: MIT Press.

Hallam, J (2002), Vocation to profession: Changing images of nursing in Britain, *Journal of Organizational Change*, Vol. 15, (1), pp. 35–47.

Hammer, E and Österle, A (2003), Welfare state policy and informal long-term care giving in Austria: Old gender divisions and new stratification processes among women, *Journal of Social Policy*, Vol. 32, (1), pp. 37–53.

Hanaoka, C and Norton, E (2008), Informal and formal care for elderly persons: How adult children's characteristics affect the use of formal care in Japan, *Social Science & Medicine*, Vol. 67, pp. 1002–1008.

Haneke, M (2012), *Amour*, Film, France/Germany/Austria.

Hanlon, N, Halseth, G, Clasby, R and Pow, V (2007), The place embeddedness of social care: Restructuring work and welfare In Mackenzie, B C (ed.), *Health & Place*, Vol. 13, pp. 466–481.

Harlton, S, Keating, N and Fast, J (1998), Defining eldercare for policy and practice: Perspectives matter, *Family Relations*, Vol. 47, (3), pp. 281–288.

Harrefors, C, Sävenstedt, S and Axelsson, K (2009), Elderly people's perceptions of how they want to be cared for: An interview study with healthy elderly couples in Northern Sweden, *Scandinavian Journal of Caring Sciences*, Vol. 23, (2), pp. 353–360.

Harris, J (2002), Caring for citizenship, *British Journal of Social Work*, Vol. 32, pp. 267–281.

Hartsock, N (1983), The Feminist Standpoint: Developing the Ground for a Specifically Feminist Historical Materialism, In Harding, S and Hintikka, B (eds.), *Discovering Reality: Feminist Perspectives on Epistemology, Metaphysics, Methodology, and Philosophy of Science*, Reidel Publishing, pp. 283–310.

Haylett, C (2003), Class, care, and welfare reform: Reading meanings, talking feelings, *Environment and Planning A*, Vol. 35, pp. 799–814.

Heaphy, B, Yip, A and Thompson, D (2004), Ageing in a non-heterosexual context, *Ageing & Society*, Vol. 24, pp. 881–902.

Heaton, J (1999), The gaze and visibility of the carer: A foucauldian analysis of the discourse of informal care, *Sociology of Health & Illness*, Vol. 21, (6), pp. 759–777.

Held, V (1990), Feminist transformations of moral theory, *Philosophy and Phenomenological Research*, Vol. 50, Suppl., pp. 321–344.

Held, V (2002a), Care and the extension of markets, *Hypatia*, Vol. 17, (2), pp. 19–33.

Held, V (2002b), Moral subjects: The natural and the normative, *Proceedings and Addresses of the American Philosophical Association*, Vol. 76, (2), pp. 7–24.

Henderson, J and Forbat, L (2002), Relationship-based social policy: Personal and policy constructions of 'care', *Critical Social Policy*, Vol. 22, (4), pp. 669–687.

Henriksson, L, Wrede, S and Burau, V (2006), Understanding professional projects in welfare service work: Revival of old professionalism? *Gender, Work and Organization*, Vol. 13, (2), pp. 174–192.

Hepworth, M (2003), Ageing Bodies: Aged by Culture, In Coupland, J and Gwyn, R (eds.), *Discourse, the Body, and Identity*, Basingstoke: MacMillan, pp. 89–106.

Hernes, H M (1987), *Welfare State and Woman Power: Essays in State Feminism*, Oslo: Norwegian University Press.

HM Government (2008), *Carers at the Heart of 21st-Century Families and Communities*, London: Department of Health.

Hochschild, A (2003a), *The Commercialization of Intimate Life: Notes from Home and Work*, London: University of California Press.

Hochschild, A (2003b), *The Managed Heart: Commercialization of Human Feeling*, 20th Anniversary Edition, London: University of California Press.

Hochschild, A (2012), *The Outsourced Self: Intimate Life in Market Times*, New York: Metropolitan Books.

Hockey, J and James, A (2003), *Social Identities Across the Life Course*, Basingstoke: Palgrave.

Holloway, M (2009), Dying old in the 21st century: A neglected issue for social work, *International Social Work*, Vol. 52, (6), pp. 713–725.

Honneth, A (1995), *The Struggle for Recognition: The Moral Grammar of Social Conflicts*, Cambridge: Polity Press.

Honneth, A (2003), Redistribution as Recognition: A Response to Nancy Fraser, In Fraser, N and Honneth, A (eds.), *Redistribution or Recognition? A Political-Philosophical Exchange*, London: Verso, pp. 110–197.

hooks, b (1990), *Yearning: Race, Gender, and Cultural Politics*, Boston: South End Press.

Horkheimer, M (1980), Materialismus und Moral, *Zeitschrift fuer Sozialforschung*, 1933, (2), Muenchen: Deutscher Taschenbuchverlag, Reprint, pp. 162–197.

Houston, S and Dolan, P (2008), Conceptualising child and family support: The contribution of honneth's critical theory of recognition, *Children & Society*, Vol. 22, (6), pp. 458–469.

Hughes, B, McKie, L, Hopkins, D and Watson, N (2005), Love's labours lost? Feminism, the disabled people's movement and an ethic of care, *Sociology*, Vol. 39, (2), pp. 259–275.

Hughes, M (2013), Decriminalising an expected death in the home: A social work response, *British Journal of Social Work*, Vol. 43, pp. 282–297.

Hugman, R (2003), Professional ethics in social work: Living with the legacy, *Australian Social Work*, Vol. 56, (1), 5–15.

Hursthouse, R (1999), *On Virtue Ethics*, Oxford: Oxford University Press.

Husso, M and Hirvonen, H (2012), Gendered agency and emotions in the field of care work, *Gender, Work and Organization*, Vol. 19, (1), pp 29–51.

Hutcheon, L (1998), *Irony, Nostalgia, and the Postmodern*, Retrieved on 15 October, 2013 from http://www.library.utoronto.ca/utel/criticism/hutchinp.html.

IDFA, *Synopsis: The Care Factory*, Retrieved on 12 May, 2014 from http://www.idfa.nl/industry/tags/project.aspx?id=56c9b901-bce8-4bd7-9dac-96f692c301da.

Industry, *Journal of Ethnic and Migration Studies*, Vol. 33, (5), pp. 801–824.

Innes, A (2009), *Dementia Studies: A Social Science Perspective*, London: Sage Publications.

Ivanhoe, P J (2007), Filial Piety as a Virtue, In Walker, R L and Ivanhoe, P J (eds.), *Working Virtue: Virtue Ethics and Contemporary Moral Problems*, Oxford: Oxford University Press, pp. 297–312.

Jäger, S (2001), Discourse and knowledge: Theoretical and Methodological Aspects of a Critical Discourse and Dispositive Analysis, In Wodak, R and Meyer, M (eds.), *Methods of Critical Discourse Analysis*, London: Sage Publications, pp. 32–62.

Jamieson, L (1998), *Intimacy: A Personal Relationship in Modern Societies*, Cambridge: Polity Press.

Jegermalm, M (2006), Informal care in Sweden: A typology of care and caregivers, *International Journal of Social Welfare*, Vol. 15, pp. 332–243.

Jenks, C (ed.) (1998), *Core Sociological Dichotomies*, London: Sage Publications.

Jenson, J and Sineau, M (2003), The Care Dimension in Welfare State Redesign, In Jenson, J and Sineau, M (eds.), *Who Cares? Women's Work, Childcare, and Welfare State Redesign*, Toronto: University of Toronto Press, pp. 3–18.

Johnson, M (1990), Dependency and Independency, In Bond, J and Coleman, P (eds.), *Aging in Society: An Introduction to Social Gerontology*, London: Sage Publications, pp. 209–228.

Johnstone, M (2013), *Alzheimer's Disease, Media Representations and the Politics of Euthanasia: Constructing Risk and Selling Death in an Ageing Society*, Furnham: Ashgate.

Jones, J (2008), 'The good old days': In-migration, social control and the decline of the 'imagined' community in a north Wales market town, *Community, Work & Family*, Vol. 11, (1), pp. 19–36.

Jones, L and Green, J (2006), Shifting discourses of professionalism: A case study of general practitioners in the United Kingdom, *Sociology of Health & Illness*, Vol. 28, (7), pp. 927–950.

Kaplan, D and Berkman, B (2011), Dementia care: A global concern and social work challenge, *International Social Work*, Vol. 54, (3), pp. 361–373.

Karlsson, G (1998), Social Democratic Women's Coup in the Swedish Parliament, In Von der Fehr, D, Rosenbeck, B and Jónasdóttir, A G (eds.), *Is There a Nordic Feminism? Nordic Feminist Thought on Culture and Society*, London: UCL Press, pp. 44–68.

Karner, C (2008), The market and the nation: Austrian (dis)agreements, *Social Identities*, Vol. 14, (2), pp. 161–187.

Karner, T X (1998), Professional caring: Homecare workers as fictive kin, *Journal of Aging Studies*, Vol. 12, (1), pp. 69–82.

Kendrick, K and Robinson, S (2002), 'Tender loving care' as a relational ethic in nursing practice, *Nursing Ethics*, Vol. 9, (3), pp. 291–300.

Kershaw, P (2005), *Carefair: Rethinking the Responsibilities and Rights of Citizenship*, Vancouver: UBC Press.

Kittay, E F (1999), *Love's Labour: Essays on Women, Equality, and Dependency*, London: Routledge.

Kittay, E F (2002), *Love's Labor* Revisited, *Hypatia*, Vol. 17, (3), pp. 237–250.

Kittay, E F (2007), Beyond Autonomy and Paternalism: The Caring Transparent Self, In Nys, T, Denier, Y and Vandevelde, T (eds.), *Autonomy & Paternalism: Reflections on the Theory and Practice of Health Care*, Leuven: Peeters Publishing, pp. 23–70.

Kittay, E F (2009), The personal is philosophical is political: A philosopher and mother of a cognitively disabled person sends notes from the battlefield, *Metaphilosophy*, Vol. 40, (3–4), pp. 606–624.

Kivisto, P and Faist, T (2007), *Citizenship: Discourse, Theory, and Transnational Prospects*, Oxford: Blackwell.

Knijn, T and Kremer, M (1997), Gender and the caring dimension of welfare states: toward inclusive citizenship, *Social Politics*, Vol. 4, (3), pp. 328–361.

Kontos, P C (1998), Resisting institutionalization: Constructing old age and negotiating home, *Journal of Ageing Studies*, Vol. 12, (2), pp. 167–184.

Kremer, M (2007), *How Welfare States Care: Culture, Gender and Parenting in Europe*, Amsterdam: Amsterdam University Press.

Krzyżanowski, M (2008), Analyzing Focus Group Discussions, In Wodak, R and Krzyżanowski, M (eds.), *Qualitative Discourse Analysis in the Social Sciences*, Basingstoke: Palgrave Macmillan, pp. 162–181.

Künemund, H (2008), Intergenerational Relations Within the Family and the State, In Saraceno C (ed.), *Families, Ageing and Social Policy*, Cheltenham: Edward Elgar, pp. 105–122.

Kvande, E and Brandth, B (2009), Gendered or gender-neutral care politics for fathers? *Annals of the American Academy of Political and Social Science*, Vol. 624, (1), pp. 177–189.

Laabs, C A (2008), The community of nursing: Moral friends, moral strangers, moral family, *Nursing Philosophy*, Vol. 9, pp. 225–232.

Landes, J B (ed.) (1998), *Feminism, the Public and the Private*, Oxford: Oxford University Press.

Larrabee, M J (1993), Gender and Moral Development: A Challenge for Feminist Theory, In Larrabee, M J (ed.), *An Ethic of Care: Feminist and Interdisciplinary Perspectives*, London: Routledge, pp. 3–16.

Latimer, J (1999), The dark at the bottom of the stairs: Performance and participation of hospitalized older people, *Medical Anthropology Quarterly*, New Series, Vol. 13, (2), pp. 186–213.

Latimer, J (2000), *The Conduct of Care: Understanding Nursing Practice*, Oxford: Blackwell Science.

Latimer, J (2012), Home Care and Frail Older People: Relational Extension and the Art of Dwelling, In Ceci, C, Björnsdóttir, K and Purkis M (eds.), *Perspectives on Care at Home for Older People*, New York: Routledge, pp. 35–61.

Lawson, V (2007), Geographies of care and responsibility, *Annals of the Association of American Geographers*, Vol. 97, (1), pp. 1–11.

Lefebvre, H (2000), *Everyday Life in the Modern World*, London: Continuum.

Leira, A and Saraceno, C (2006), Care: Actors, relationships, contexts, *Sosiologi I Dag*, Vol. 36, (3), pp. 7–34.

Leitner, S (2003), Varieties of familialism: The caring function of the family in comparative perspective, *European Societies*, Vol. 5, pp. 353–375.

Leonard, S and Tronto, J (2007), The genders of citizenship, *American Political Science Review*, Vol. 101, (1), pp. 33–46.

Levinas, E (2001), *Is It Righteous to Be? Interviews with Emmanuel Levinas*, Edited by Jill Robbins, Stanford, California: Stanford University Press.

Lewis, J (2007), Gender, ageing and the 'New social settlement': The importance of developing a holistic approach to care policies, *Current Sociology*, Vol. 55, (2), pp. 271–286.

Liang, J and Luo, B (2012), Toward a discourse shift in social gerontology: From successful aging to harmonious aging, *Journal of Aging Studies*, Vol. 26, pp. 327–334.

Liaschenko, J and Peter, E (2004), Nursing ethics and conceptualizations of nursing: Profession, practice and work, *Journal of Advanced Nursing*, Vol. 46, (5), pp. 488–495.

Lister, R (1997), *Citizenship: Feminist Perspectives*, Basingstoke: MacMillan.

Lister, R, Williams, F, Anttonen, A, Bussemaker, J, Gerhard, U, Heinen, J, Johansson, S and Leira, A (2007), *Gendering Citizenship in Western Europe: New Challenges for Citizenship Research in a Cross-National Context*, Bristol: The Policy Press.

Lloyd, L (2004), Mortality and morality: Ageing and the ethics of care, *Ageing & Society*, Vol. 24, pp. 235–256.

Lloyd, L (2006), Call us carers: Limitations and risks in campaigning for recognition and exclusivity, *Critical Social Policy*, Vol. 26, (4), pp. 945–960.

López, D and Domènech, M (2009), Embodying autonomy in a home telecare service, *The Sociological Review*, Vol. 56, (s2), pp. 181–195.

Lukes, S (1973), *Emile Durkheim: His Life and Work: A Historical and Critical Study*, Stanford: Stanford University Press.

Lynch, K (2007), Love labour as a distinct and non-commodifiable form of care labour, *Sociological Review*, Vol. 55, (3), pp. 550–570.

Lyon, D and Glucksmann, M (2008), Comparative configurations of care work across Europe, *Sociology*, Vol. 42, (1), pp. 101–118.

Macdonald, C L and Merrill D A (2002), 'It shouldn't have to be A trade': Recognition and redistribution in care work advocacy, *Hypatia*, Vol. 17, (2), pp. 67–83.

Mallett, S (2004), Understanding home: A critical review of the literature, *Sociological Review*, Vol. 52, (1), pp. 62–89.

Malone, M and Dooley, J (2006), 'Dwelling in displacement': Meanings of 'community and *sense* of community for two generations of Irish people living in North-West London, *Community, Work and Family*, Vol. 9, (1), pp. 11–28.

Mand, K (2006), Social Relations beyond the family? Exploring elderly South Asian women's friendships in London, *Community, Work and Family*, Vol. 9, (3), pp. 309–323.

Manthorpe, J and Iliffe, S (2009), Changing the culture of social work support for people with early dementia, *Australian Social Work*, Vol. 62, (2), pp. 232–244.

Martin-Matthews, A (2007), Situating 'Home' at the nexus of the public and private spheres: Ageing, gender and home support work in Canada, *Current Sociology*, Vol. 55, (2), pp. 229–249.

Massey, D (1995), The Conceptualization of Place, In Massey, D and Jess, P (eds.), *A Place in the World? Places, Cultures and Globalization*, Oxford: Oxford University Press, pp. 215–239.

Massey, D (2007), *World City*, Cambridge: Polity Press.

Mautner, G (2008), Analyzing Newspapers, Magazines and Other Print Media, In Wodak, R and Krzyżanowski, M (eds.), *Qualitative Discourse Analysis in the Social Sciences*, Basingstoke: Palgrave Macmillan, pp. 30–53.

McDonald, R, Warring, J and Harrison, S (2006), At the cutting edge? Modernization and nostalgia in a hospital operating theatre department, *Sociology*, Vol. 40, (6), pp. 1097–1115.

McDowell, L (1999), *Gender, Identity and Place: Understanding Feminist Geographies*, Cambridge: Polity Press.

McGregor, J (2007), *Joining the BBC (British Bottom Cleaners)*: Zimbabwean Migrants and the UK care industry, *Journal of Ethnic and Migration Studies*, Vol. 33, (5), pp. 801–824.

McKie, L, Bowlby, S and Gregory, S (2001), Gender, caring and employment in Britain, *Journal of Social Policy*, Vol. 30, (2), pp. 233–258.

McNay, L (1994), *Foucault: A Critical Introduction*, Cambridge: Polity Press.

Meagher, G and Szebehely, M (eds.) (2013), *Marketisation in Nordic Elder Care: A Research Report on Legislation, Oversight, Extent and Consequences*, Stockholm:

Stockholm University, http://www.normacare.net/wp-content/uploads/2013/09/Marketisation-in-nordic-eldercare-webbversion-med-omslag1.pdf, accessed July 2014.

Mee, K (2009), A space to care, a space of care: Public housing, belonging, and care in inner Newcastle, Australia, *Environment and Planning A*, Vol. 41, pp. 842–858.

Mehta, K K and Thang, L L (2008), Visible and Blurred Boundaries in Familial Care: The Dynamics of Multigenerational Care in Singapore, In Martin-Matthews, A and Phillips, J E (eds.), *Aging and Caring at the Intersection of Work and Home Life: Blurring the Boundaries*, Hove: Psychology Press, pp. 43–83.

Meis, C de, Almeida Souza, C de and Sivla Filho, J de (2007), House and street: Narratives of professional identity among nurses, *Journal of Professional Nursing*, Vol. 23, (6), pp. 325–328.

Milligan, C (2003), Location or dis-location? Towards a conceptualization of people and place in the care-giving experience, *Social & Cultural Geography*, Vol. 4, (4), pp. 455–470.

Milligan, C (2005), From home to 'home': Situating emotions within the caregiving experience, *Environment and Planning A*, Vol. 37, pp. 2105–2120.

Mills, C. (2000), *The Sociological Imagination*, 40th ed., Oxford: Oxford University Press.

Misra, J, Moller, S and Karides, M (2003), Envisioning dependency: Changing media depictions of welfare in the 20th Century, *Social Problems*, Vol. 50, (4), pp. 482–504.

Mittelstadt, J (2001), 'Dependency as a problem to be solved': Rehabilitation and the American liberal consensus on welfare in the 1950s, *Social Politics*, Vol. 8, (2), pp. 228–257.

Mitteness, L and Barker, J (1995), Stigmatising a 'Normal' condition: Urinary incontinence in late life, *Medical Anthropology Quarterly*, New Series, Vol. 9, (2), pp. 188–210.

Moll, A (2002), *The Body Multiple: Ontology in Medical Practice*, Durham: Duke University Press.

Morgan, D (1996), *Family Connections*, Cambridge: Polity Press.

Morgan, K and Zippel, J (2003), Paid to care: The origins and effects of care leave policies in western Europe, *Social Politics*, Vol. 10, (1), pp. 49–85.

Mullen, P (2002), *The Imaginary Time Bomb: Why an Ageing Population Is Not a Social Problem*, London: Tauris.

Munro, R and Belova, O (2009), The body in time: Knowing bodies and the 'interruption' of narrative, *The Sociological Review*, Vol. 56, (s2), pp. 87–99.

Muzio, D and Tomlinson, J (2012), Editorial: Researching gender, inclusion and diversity in contemporary professions and professional organizations, *Gender, Work and Organization*, Vol. 19, (5), pp. 455–466.

Nafstad, H E, Carlquist, E and Blakar, R M (2007), Community and care work in a world of changing ideologies, *Community, Work & Family*, Vol. 10, (3), pp. 329–340.

Nelson, J A and England, P (2002), Feminist philosophies of love and work, *Hypatia*, Vol. 17, (2), pp. 1–18.

Nisbet, R A (1966), *The Sociological Tradition*, London: Heinemann.

Noddings, N (2002), *Starting at Home*, California: University of California Press.

Noddings, N (2003), *Caring: A Feminine Approach to Ethics and Moral Education*, 2nd ed., London: University of California Press.

Nussbaum, M (2002), The Future of Feminist Liberalism, In Kittay, E F and Feder, E K (eds.), *The Subject of Care: Feminist Perspectives on Dependency*, Lanham: Rowner & Littlefield, pp. 186–214.

Nussbaum, M (2006), *Frontiers of Justice: Disability, Nationality, Species Membership*, London: The Belknap Press.

O'Neill, F (2009), Bodily knowing as uncannily canny: Clinical and ethical significance, *The Sociological Review*, Vol. 56, (s2), pp. 216–232.

Oakley, A (1974), *The Sociology of Housework*, New York: Pantheon Books.

Oakley, A (2007), *Fracture: Adventures of a Broken Body*, Bristol: The Policy Press.

Oldman, C (2003), Deceiving, theorizing and self-justification: A critique of independent living, *Critical Social Policy*, Vol. 23, (1), pp. 44–62.

Oliver, M (1990), *The Politics of Disablement*, Basingstoke: Macmillan.

Oliver, M and Barnes, C (1998), *Disabled People and Social Policy: From Exclusion to Inclusion*, London: Longman.

ONS (2013), More than 1 in 10 providing unpaid care as numbers rise to 5.8 million, Office for National Statistics: http://www.ons.gov.uk/ons/rel/mro/news-release/how-much-unpaid-care-are-the-residents-of-england-and-wales-providing-in-2011/unpaid-care-in-england-and-wales.html, accessed July 2014.

Orloff, A (1993), Gender and the social rights of citizenship: The comparative analysis of gender relations and welfare states, *American Sociological Review*, Vol. 58, (3), pp. 303–328.

Österle, A (2001), *Equity Choices and Long-Term Care Policies in Europe: Allocating Resources and Burdens in Austria, Italy, the Netherlands and the UK*, Aldershot: Ashgate Publishers.

Österle, A and Bauer, G (2012), Home care in Austria: The interplay of family orientation, cash-for-care and migrant care, *Health and Social Care in the community*, Vol. 20, pp. 265–273.

Österle, A and Hammer, E (2004), *Zur zukünftigen Betreuung und Pflege älterer Menschen: Rahmenbedingungen – Politikansätze – Entwicklungsperspektiven*, on behalf of Caritas Österreich, Vienna: Kardinal König Akademie.

Outshoorn, J (2002), Gendering the 'greying' of society: A discourse analysis of the care gap, *Public Administration Review*, Vol. 62, (2), pp. 185–196.

Özyürek, E (2006), *Nostalgia for the Modern: State Secularism and Everyday Politics in Turkey*, Durham: Duke University Press.

Paoletti, I (2001), Membership categories and time appraisal in interviews with family caregivers of disabled elderly, *Human Studies*, Vol. 24, (4), pp. 293–325.

Paoletti, I (2002), Caring for older people: A gendered practice, *Discourse & Society*, Vol. 13, (6), pp. 805–817.

Parker, M (ed.) (1999), *Ethics and Community in the Health Care Professions*, London: Routledge.

Parks, J A (2002), *No Place Like Home? Feminist Ethics and Home Health Care*, Bloomington: Indiana University Press.

Parr, H and Philo, C (2003), Rural mental health and social geographies of caring, *Social and Cultural Geography*, Vol. 4, (4), pp. 471–488.

Pfau-Effinger, B (2005), Welfare state policies and the development of care arrangements, *European Societies*, Vol. 7, (2), pp. 321–347.

Pfau-Effinger, B and Geissler, B (eds.) (2005), *Care Arrangements in Europe – Variations and Change*, Bristol: The Policy Press.

Phillips J E and Bernard M (2008), Work and Care: Blurring the Boundaries of Space, Place, Time, and Distance, In Martin-Matthews, A and Phillips J E (eds.), *Aging and Caring at the Intersection of Work and Home Life: Blurring the Boundaries*, New York/Hove: Psychology Press, pp. 85–105.

Phillips, J (2007), *Care, Key Concepts*, Cambridge: Polity Press.

Pickard, S (2010), The 'Good carer': Moral practices in late modernity, *Sociology*, Vol. 44, (3), pp. 471–487.

Pickering, M and Keightley, E (2006), The modalities of nostalgia, *Current Sociology*, Vol. 54, (6), pp. 919–941.

Pink, S (2004), *Home Truths: Gender, Domestic Objects and Everyday Life*, Oxford: Berg Publishers.

Plath, D (2008), Independence in old age: The route to social exclusion? *British Journal of Social Work*, Vol. 38, pp. 1353–1369.

Powell, J (2006), *Social Theory and Aging*, Lanham: Rowman & Littlefield.

Pourtova, E (2013), Nostalgia and lost identity, *Journal of Analytical Psychology*, Vol. 58, (1), pp. 34–51.

Power, A (2008), Caring for independent lives: Geographies of caring for young adults with intellectual disabilities, *Social Science & Medicine*, Vol. 67, pp. 834–843.

Price, E (2008), Pride or prejudice? Gay men, lesbians and dementia, *British Journal of Social Work*, Vol. 38, pp. 1337–1352.

Qureshi, H and Walker, A (1989), *The Caring Relationship: Elderly People and Their Families*, Basingstoke: Macmillan.

Reisigl, M and Wodak, R (2001), *Discourse and Discrimination: Rhetorics of Racism and Antisemitism*, London: Routledge.

Robertson, R (1992), *Globalization: Social Theory and Global Culture*, London: Sage Publications.

Robertson, R (1995), Glocalization: Time-Space and Homogeneity-Heterogeneity, In Featherstone, M, Lash, S and Robertson, R (eds.), *Global Modernities*, London: Sage Publications, pp. 25–44.

Roos, P, Trigg, M and Hartman, M (2006), Changing families/changing communities: work, family and community in transition, *Community, Work and Family*, Vol. 9, (2), pp. 197–224.

Rose, H (1983), Hand, brain, and heart: A feminist epistemology for the natural sciences, *Signs*, Vol. 9, (1), pp. 73–90.

Rosenbeck, B (1998), Nordic Women's Studies and Gender Research, In Von der Fehr, D, Rosenbeck, B and Jónasdóttir, A G (eds.), *Is There a Nordic Feminism? Nordic Feminist Thought on Culture and Society*, London: UCL Press, pp. 344–357.

Roseneil, S and Budgeon, S (2004), Cultures of intimacy and care beyond 'the family': Personal life and social change in the early 21st century, *Current Sociology*, Vol. 52, (2), pp. 135–159.

Rosenthal, C, Martin-Matthews, A and Keefe, J (2007), Care management and care provision for older relatives amongst employed informal care-givers, *Ageing & Society*, Vol. 27, pp. 755–778.

Rozanova, J (2010), Discourse of successful aging in *The Globe & Mail. Insights from Critical Gerontology*, Vol. 24, pp. 213–222.

Rubenstein, R (2001), *Home Matters: Longing and Belonging, Nostalgia and Mourning in Women's Fiction*, New York: Palgrave.

Ruddick, S (2002), An appreciation of *Love's Labor*, *Hypatia*, Vol. 17, (3), pp. 214–224.

Rudge, T (2009), Beyond caring? Discounting the differently known body, *The Sociological Review*, Vol. 56, (s2), pp. 233–248.

Rudman, D (2006), Shaping the active, autonomous and responsible modern retiree: an analysis of discursive technologies and their links with neo-liberal political rationality, *Ageing & Society*, Vol. 26, pp. 181–201.

Rummery, K (2009), A Comparative discussion of the gendered implications of cash-for-care schemes: Markets, independence and social citizenship in crisis? *Social Policy & Administration*, Vol. 43, (6), pp. 634–648.

Russell, C (2007), What do older women and men want? Gender differences in the 'lived experience' of ageing, *Current Sociology*, Vol. 55, (2), pp. 173–192.

Ryburn, B, Wells, Y and Foreman, P (2009), Enabling independence: Restorative approaches to home care provision for frail older adults, *Health and Social Care in the Community*, Vol. 17, (3), pp. 225–234.

Said, E (2003), *Orientalism*, London: Penguin.

Sainsbury, D (1996), *Gender, Equality, and the Welfare States*, Cambridge: Cambridge University Press.

Saussure, F de (1972), *Course in General Linguistics*, Peru: Open Court Publishing.

Sayer, A (2011), *Why Things Matter to People: Social Science, Values and Ethical Life*, Cambridge: Cambridge University Press.

Schmid Noerr, G (1997), *Gesten aus Begriffen: Konstellationen der Kritischen Theorie*, Frankfurt am Main: Fischer.

Schmid, T, Brandt, M and Haberkern, K (2012), Gendered support to older parents: do welfare states matter? *European Journal of Ageing*, Vol. 9, pp. 39–50.

Scott, J W (2004), Feminist Family Politics, In Scott, J W and Keates D (eds.), *Going Public: Feminism and the Shifting Boundaries of the Private Sphere*, Urbana: University of Illinois Press, pp. 225–237.

Scott, J W and Keates D (eds.) (2004), *Going Public: Feminism and the Shifting Boundaries of the Private Sphere*, Urbana: University Of Illinois Press.

Scourfield, P (2006), 'What matters is what works?' How discourses of modernization have both silenced and limited debate on domiciliary care for older people, *Critical Social Policy*, Vol. 26, (1), pp. 5–30.

Sennett, R (2003), *Respect: The Formation of Character in a World of Inequality*, London: Penguin Press.

Sevenhuijsen, S (1998), *Citizenship and the Ethics of Care: Feminist Considerations on Justice, Morality and Politics*, London: Routledge.

Shakespeare, T (2000), *Help*, Birmingham: Venture Press.

Shilling, C (1993), *The Body and Social Theory*, London: Sage Publications.

Shilling, C (1997), The undersocialised conception of the (embodied) agent in modern sociology, *Sociology*, Vol. 31, (4), pp. 737–754.

Shutes, I (2012), The employment of migrant workers in long-term care: Dynamics of choice and control, *Journal of Social Policy*, Vol. 41, (1), pp. 43–59.

Siim, B (1987), The scandinavian welfare states – towards sexual equality or a new kind of male domination? *Acta Sociologica*, Vol. 30, (3/4), pp. 255–270.

Siim, B (1993), The Gendered Scandinavian Welfare States: The Interplay between Women's Roles as Mothers, Workers and Citizens in Denmark, In Lewis, J (ed.), *Women and Social Policies in Europe: Work, Family and the State*, Aldershot: Edward Elgar Publishing, pp. 25–48.

Silva, E B and Smart C (1999), The 'New' Practices and Politics of Family Life, In Silva, E B and Smart, C (eds.), *The New Family?* London: Sage Publications, pp. 1–12.

Simonazzi A (2009), Care regimes and national employment models, *Cambridge Journal of Economics*, Vol. 33, (2), pp. 211–232.

Skjeie, H and Siim, B (2000), Scandinavian feminist debates on citizenship, *International Political Science Review*, Vol. 21, (4), pp. 435–460.

Slote, M (2001), *Morals from Motives*, Oxford: Oxford University Press.

Smart, B (1999), *Facing Modernity: Ambivalence, Reflexivity and Morality*, London: Sage Publications.

Smart, C (2007), *Personal Life: New Directions in Sociological Thinking*, Cambridge: Polity Press.

Smart, C and Neale, B (1999), *Family Fragments?* Cambridge: Polity Press.

Smith, R, Heley, J and Stafford, I (2011), Woolworths and wales: A multidimensional analysis of the loss of a local brand, *Sociological Research Online*, Vol. 16, (1), DOI: 10.5153/sro.2284.

Smith, S R (2005), Equality, identity and the disability rights movement: From policy to practice and from Kant to Nietzsche in more than one uneasy move, *Critical Social Policy*, Vol. 25, (4), pp. 554–576.

Sörensen, K and Bergqvist, C (2002), *Gender and the Social Democratic Welfare Regime: A Comparison of Gender-Equality Friendly Policies in Sweden and Norway*, Work Life in Transition, Vol. 5, Stockholm: National Institute for Working Life.

Sprigings, N and Allen, C (2005), The communities we are regaining but need to lose: A critical commentary on community building in beyond-place societies, *Community, Work & Family*, Vol. 8, (4), pp. 389–411.

Stackhouse, J (1998), *Into the Community: Nursing in Ambulatory and Home Care*, Philadelphia: Lippincott.

Staeheli, L and Brown, M (2003), Guest editorial, *Environment and Planning A*, Vol. 35, pp. 771–777.

Statistik Austria (2014a), Bundespflegegeldbezieherinnen und -bezieher sowie Ausgaben für das Bundespflegegeld 1993–2012, http://www.statistik.at/web_de/statistiken/soziales/sozialleistungen_auf_bundesebene/bundespflegegeld/020069.html, accessed July 2014.

Statistik Austria (2014b), Landespflegegeldbezieherinnen und -bezieher sowie Ausgaben für das Landespflegegeld 2001–2011, http://www.statistik.at/web_de/statistiken/soziales/sozialleistungen_auf_landesebene/landespflegegeld/020136.html, accessed July 2014.

Still, J (1994), 'What Foucault fails to acknowledge ...', feminists and the history of sexuality, *History of the Human Sciences*, Vol. 7, (2), pp. 150–157.

Strangleman, T (2012), Work identity in crisis? Rethinking the problem of attachment and loss at work, *Sociology*, Vol. 46, (3), pp. 411–425.

Strathern, M (1997), Gender: Division or Comparison, In Hetherington, K and Munro, R (eds.), *Ideas of Difference: Social Spaces and the Labour of Division*, Sociological Review Monograph, Oxford: Blackwells.

Svašek, M (2008), Who cares? Families and feelings in movement, *Journal of Intercultural Studies*, Vol. 29, (3), pp. 213–230.

Swenson, C (2004), Dementia diary: A personal and professional Journal, *Social Work*, Vol. 49, (3), pp. 451–460.

Tanner, D (2013), Identity, selfhood and dementia: Messages for social work, *European Journal of Social Work*, Vol. 16, (2), pp. 155–170.

Tate, S (2007), Translating melancholia: A poetics of black interstitial community, *Community, Work and Family*, Vol. 10, (1), pp. 1–15.

Taylor, D (1998), Social identity and social policy: Engagements with postmodern theory, *Journal of Social Policy*, Vol. 27, (3), pp. 329–350.

Thien, D and Hanlon, N (2009), Unfolding dialogues about gender, care and 'the north': An introduction, *Gender, Place & Culture*, Vol. 16, (2), pp. 155–162.

Thompson, E P (1991), *Customs in Common*, London: The Merlin Press.

Timonen, V and Doyle, M (2007), Worlds apart? Public, private and non-profit sector providers of domiciliary care for older persons in Ireland, *Journal of Aging Studies*, Vol. 21, pp. 255–265.

Todorova, M and Gille, Z (eds.) (2010), *Post-Communist Nostalgia*, Oxford: Berghahn Books.

Toffanin, T (2011), The role of neo-liberal capitalism in reproducing gender inequality in Italy, *Journal of Contemporary European Studies*, Vol. 19, pp. 379–393.

Tönnies, F (1955), *Community and Association (Gemeinschaft und Gesellschaft)*, London: Routlegdge.

Townsend, P (1955), The family life of old people: An investigation in East London, *The Sociological Review*, Vol. 3, (2), pp. 175–195.

Townsend, P (1962), *The Last Refuge: A Survey of Residential Institutions and Homes for the Aged in England and Wales*, Routledge and Kegan Paul.

Tronto, J (1993), *Moral Boundaries: A Political Argument for an Ethic of Care*, London: Routledge.

Tronto, J (2013), *Caring Democracy: Markets, Equality, and Justice*, New York: New York University Press.

Tronto, J and Weicht, B (forthcoming), 'As Long as Care Is Attached to Gender There Is No Justice', Interview with Joan C. Tronto, *Tijdschrift voor Genderstudies*

Twigg, J (1997), Deconstructing the 'social bath': Help with bathing at home for older and disabled people, *Journal of Social Policy*, Vol. 26, pp. 211–232.

Twigg, J (2000a), *Bathing – the Body and Community Care*, London: Routledge.

Twigg, J (2000b), Carework as a form of bodywork, *Ageing & Society*, Vol. 20, pp. 389–411.

Twigg, J (2003), Social Care, In Baldock, J, Manning, N and Vickerstaff, S (eds.), *Social Policy*, 2nd ed., Oxford: University Press, pp. 419–453.

Ungerson, C (1987), *Policy Is Personal: Sex, Gender, and Informal Care*, London: Tavistock Publications.

Ungerson, C (1997), Social politics and the commodification of care, *Social Politics*, Vol. 4, (3), pp. 362–381.

Ungerson, C (1999), Personal assistants and disabled people: An examination of a hybrid form of work and care, *Work, Employment & Society*, Vol. 13, (4), pp. 583–600.

Ungerson, C (2000), Thinking about the production and consumption of long-term care in Britain: Does gender still matter? *Journal of Social Policy*, Vol. 29, (4), pp. 623–643.

Ungerson, C (2004), Whose empowerment and independence? A cross-national perspective on 'cash for care' schemes, *Ageing and Society*, Vol. 24, (2), pp. 189–212.

Ungerson, C (2005), Care, work and feeling, *The Sociological Review*, Vol. 53, (2), pp. 188–203.

Vabø, M and Szebehely, M (2012), A Caring State for All Older People? In Anttonen, A, Häikiö, L and Stefánsson, K (eds.), *Welfare State, Universalism and Diversity*, Cheltenham: Edward Elgar, pp. 121–144.

Valverde, M (2004), Experience and Truth Telling in a Post-Humanist World: A Foucauldian Contribution to Feminist Ethical Reflections, In Taylor, D and Vintges, K (eds.), *Feminism and the Final Foucault*, Urbana and Chicago: University of Illinois Press, pp. 67–90.

van der Geest, S, Mul A and Vermeulen, H (2004), Linkages between migration and the care of frail older people: Observations from Greece, Ghana and The Netherlands, *Ageing & Society*, Vol. 24, pp. 431–450.

van Dijk, T A (1988), *News Analysis: Case Studies of International and National News in the Press*, Hillsdale/New Jersey: Lawrence Erlbaum Assoc.

van Dijk, T A (1991), *Racism and the Press*, London: Routledge.

van Leeuwen, T (1995), Representing social action, *Discourse & Society*, Vol. 6, (1), pp. 81–106.

Varley, A (2008), A place like this? Stories of dementia, home, and the self, *Environment and Planning D: Society and Space*, Vol. 26, pp. 47–67.

Verhulst, D (2013), *De Laatkomer*, Amersfoort: Atlas Contact.

Wærness, K (1998), The Changing 'Welfare Mix' in Childcare and Care for the Frail Elderly in Norway, In Lewis, J (ed.), *Women and Social Policies in Europe: Work, Family and the State*, Aldershot: Ashgate Publishing, pp. 207–228.

Walker, A (1995), The Family and the Mixed Economy of Care – Can They be Integrated? In Allen, I and Perkins, E (eds.), *The Future of Family Care for Older People*, London: HMSO, pp. 201–218.

Watson, N, McKie, L, Hughes, B, Hopkins, D and Gregory, S (2004), (Inter)Dependence, needs and care: The potential for disability and feminist theorists to develop and emancipatory model, *Sociology*, Vol. 38, (2), pp. 331–350.

Watson, S and Wells, K (2005), Spaces of nostalgia: The hollowing out of a London market, *Social & Cultural Geography*, Vol. 6, (1), pp. 17–30.

Weicht, B (2010), Embodying the ideal carer: The Austrian discourse on migrant carers, *International Journal of Ageing and Later Life*, Vol. 5, (2), pp. 17–52.

Weicht, B (2011), Embracing dependency: Rethinking (in)dependence in the discourse of care, *The Sociological Review*, Vol. 58, (s2), pp. 205–224.

Weicht, B (2013), The making of 'the elderly': Constructing the subject of care, *Journal of Aging Studies*, Vol. 27, pp. 188–197.

Weicht, B (2015), Employment Without Employers? The Public Discourse on Care During the Regularisation Reform in Austria, In Triandafyllidou, A and Marchetti, S (eds.), *Employers, Agencies and Immigration: Paying for Care*, Farnham: Ashgate Publishing, pp. 113–130.

Wetherell, M and Potter, J (1992), *Mapping the Language of Racism: Discourse and the Legitimation of Exploitation*, Hemel Hempstead: Harvester Wheatsheaf.

Whitaker, A (2010), The body as existential midpoint – the aging and dying body of nursing home residents, *Journal of Aging Studies*, Vol. 24, pp. 96–104.

White, K (2002), Nursing as vocation, *Nursing Ethics*, Vol. 9, (3), pp. 279–290.

White, P (2009), Knowing body, knowing other: Cultural materials and intensive care, *The Sociological Review*, Vol. 56, (s2), pp. 117–37.

Wiles, J (2003), Daily geographies of caregivers: Mobility, routine, scale, *Social Science & Medicine*, Vol. 57, pp. 1307–1325.

Wilińska, M (2010), Because women will always be women and men are just getting older: Intersecting discourses of ageing and gender, *Current Sociology*, Vol. 58, (6), pp. 879–896.

Wilińska, M and Cedersund, E (2010), 'Classic ageism' or 'brutal economy'? Old age and older people in the polish media, *Journal of Aging Studies*, Vol. 24, (4), pp. 335–343.

Wilińska, M and Henning, C (2011), Old age identity in social welfare practice, *Qualitative Social Work*, Vol. 10, (3), pp. 346–363.

Wilkinson, S (2004), Focus Group Research, In Silverman, D (ed.), *Qualitative Research: Theory, Method, and Practice*, London: Sage Publications, pp. 177–199.

Williams, A (2002), Changing geographies of care: Employing the concept of therapeutic landscapes as a framework in examining home space, *Social Science & Medicine*, Vol. 55, pp. 141–154.

Williams, A and Crooks, V (2008), Introduction: Space, place and the geographies of women's caregiving work, *Gender, Place & Culture*, Vol. 15, (3), pp. 243–247.

Williams, F (1989), *Social Policy: A Critical Introduction: Issues of Race, Gender and Class*, Cambridge: Polity Press.

Williams, F (1996), Postmodernism, Feminism and the Question of Difference, In Parton, N (ed.), *Social Theory, Social Change and Social Work*, London: Routledge, pp. 61–76.

Williams, F (2001), In and beyond new labour: Towards a new political ethic of care, *Critical Social Policy*, Vol. 21, (4), pp. 467–493.

Williams, F (2004), *Rethinking Families*, ESRC CAVA Research Group, London: Calouste Gulbenkian Foundation.

Williams, J (2000), *Unbending Gender: Why Family and Work Conflict and What to Do About It*, Oxford: Oxford University Press.

Williams, R (1973), *The Country and the City*, London: Chatto & Windus.

Williams, R (1980), *Problems in Materialism and Culture: Selected Essays*, London: Verso.

Winch, S (2006), Constructing a morality of caring: Codes and values in Australian carer discourse, *Nursing Ethics*, Vol. 13, (1), pp. 5–16.

Wodak, R (2001a), What CDA Is About – A Summary of Its History, Important Concepts and Its Developments, In Wodak, R and Meyer, M (eds.), *Methods of Critical Discourse Analysis*, London: Sage Publications, pp. 1–13.

Wodak, R (2001b), The Discourse-Historical Approach, In Wodak, R and Meyer, M (eds.), *Methods of Critical Discourse Analysis*, London: Sage Publications, pp. 63–94.

Wodak, R (2008), Introduction: Discourse Studies – Important Concepts and Terms, In Wodak, R and Krzyżanowski, M (eds.), *Qualitative Discourse Analysis in the Social Sciences*, Basingstoke: Palgrave Macmillan, pp. 1–29.

Wodak, R, de Cillia, R, Cinar, D and Matouschek, B (1995), Identitätswandel Österreichs im veränderten Europa: Diskurshistorische Studien über den öffentlichen und privaten Diskurs zur ‚neuen' österreischischen Identität, in Projekt-Team „Identitätswandel Österreichs im veränderten Europa" (eds.), *Nationale und kulturelle Identitäten Österreichs: Theorien, Methoden und Probleme der Forschung zu kollektiver Identität*, Vienna: IFK.

Wong, W and Ussher, J (2009), Bereaved informal cancer carers making sense of their palliative care experiences at home, *Health and Social Care in the Community*, Vol. 17, (3), pp. 274–282.

Yantzi, N and Rosenberg, M (2008), The contested meanings of home for women caring for children with long-term care needs in Ontario, Canada, *Gender, Place & Culture*, Vol. 15, (3), pp. 301–315.

Young, I M (2002), Autonomy, Welfare Reform, and Meaningful Work, In Kittay, E F and Feder, E K (eds.), *The Subject of Care: Feminist Perspectives on Dependency*, Lanham: Rowner & Littlefield, pp. 40–60.

Young, I M (2005a), A Room of One's Own: Old Age, Extended Care, and Privacy', In Young, I M, *On Female Body Experience: 'Throwing Like a Girl' and Other Essays*, Oxford: Oxford University Press, pp. 155–170.

Young, I M (2005b), House and Home: Feminist Variations on a Theme, In Young, I M, *On Female Body Experience: 'Throwing Like a Girl' and Other Essays*, Oxford: Oxford University Press, pp. 123–154.

Zadoroznyi, M (2009), Professionals, carers or 'Strangers'? Liminality and the typification of postnatal home care workers, *Sociology*, Vol. 43, (2), pp. 268–285.

Zechner, M and Sointu, L (2008), Kin, Care and Elderly Persons in Finland and Italy, Paper Presented at the *4th Congress of the European Society on Family Relations*, September 2008, University of Jyväskylä, Finland.

Zelizer, V A (2005), *The Purchase of Intimacy*, Princeton: Princeton University Press.

Zhou, X, Wildschut, T, Sedikides, C, Chen, X and Vingerhoets, A (2012), Heartwarming memories: Nostalgia maintains physiological comfort, *Emotion*, Vol. 12, (4), pp. 678–684.

Žižek, S (2000), *Das Fragile Absolute: Warum es Sich Lohnt, Das Christliche Erbe zu Verteidigen*, Berlin: Volk & Welt.

Index